Convergence of Communications, Navigation, Sensing and Services

RIVER PUBLISHERS SERIES IN COMMUNICATIONS

This includes the theory and use of systems involving all terminals, computers, and information processors; wired and wireless networks; and network layouts, procontentsols, architectures, and implementations.

Furthermore, developments toward new market demands in systems, products, and technologies such as personal communications services, multimedia systems, enterprise networks, and optical communications systems.

- Wireless Communications
- Networks
- Security
- Antennas & Propagation
- Microwaves
- Software Defined Radio

For a list of other books in this series, visit www.riverpublishers.com

Convergence of Communications, Navigation, Sensing and Services

Editors

Leo Ligthart

Chairman CONASENSE
the Netherlands

Ramjee Prasad

CTIF
Aalborg University
Denmark

River Publishers
Aalborg

Conasense

Published, sold and distributed by:
River Publishers
Niels Jernes Vej 10
9220 Aalborg Ø
Denmark

ISBN: 978-87-93102-75-0 (Print)
978-87-93102-76-7 (Ebook)

Contents

Preface

The CONASENSE foundation was established as brain tank in November 2012. Main aim is to define and steer processes directed towards actions on investigations, developments and demonstrations of novel CONASENSE services with high potential and importance for society. Characteristic for realizing the services is that integration of communications, navigation and sensing technology is needed. The horizon for new services is 2020 and beyond. Knowing that CONASENSE can play a role in a wide range of areas we decided to limit ourselves in the initial phase 2012–2015 to 2 areas with 2 respective working groups :

- Quality of life (QoL)
- Integrated CONASENSE Architectures (ICA)

Most members of the working group are connected with academia, but all have strong links with non-academic organizations, governmental and non-governmental institutes as well as industries. Some members come from semi-governmental organizations and industry. It is foreseen that the number of participating organizations will expand in 2014 and beyond.

The year 2013 was very successful for CONASENSE. We published the 1st CONASENSE book.Several articles for the CONASENSE journal have been written, reviewed and edited. Another highlight was the 2nd workshop in March 2013, followed by intensive discussions on the CONASENSE essence, that is:

- Giving contents to interconnections between communications, navigation, sensing and services
- Stimulating cooperation with multi-disciplinary groups active in technology and in development of services as well as for user groups implementing and evaluating new services
- Developing roadmaps for novel services
- Preparing proposals for activities which have impact on governments and decision makers

CONASENSE was introduced in USA, China and Indonesia.

This 2nd CONASENSE book is based on the brainstorms during the workshop in March 2012 followed by communications between authors and editors. At the occasion of publishing the book I thank firstly the authors of the book chapters. Secondly I thank the participants in the working groups for their input and discussions on the essence of CONASENSE. Special thanks go to Silvano Pupolin for his duties in the QoL working group and to Mehmet Safak and Ernestina Cianca for their efforts in the ICA working group. This 2nd CONASENSE book reflects some most interesting examples of QoL and ICA activities which will play a role in future CONASENSE initiatives.

I conclude with the same sentence written in my preface of the 1st CONASENSE book: "I hope that readers of the book are inspired by the topics that need future research and development and that they become motivated to work on those CONASENSE topics and finally that they are eager to join CONASENSE community".

Leo Ligthart
Chairman CONASENSE
www.conasense.org

1

Vision on CONASENSE Architecture

M. Şafak,[1] H. Nikookar[2] and L. Ligthart[3]

[1]University Hacettepe, Turkey
[2]Delft University of Technology, The Netherlands
[3]Chairman Conasense, The Netherlands

1.1 Introduction

Recent advances in digital communications and high-speed digital signal processing led to innovative technologies, techniques, systems and services in the areas of communications, navigation and sensing. Supported by the integration of transmission of voice, data and video using Internet Protocol (IP) and the accompanying increase in the demand, these changes greatly improved the versatility, availability and ubiquitous use of these services. Nowadays, we observe a rapidly increasing demand and innovative application areas for services related to positioning, tracking and navigation of some users/platforms. For example, we currently use available services for determining and tracking the position of a user in mountainous/forest areas or in seas, for finding the address of a colleague that we want to visit as we drive in a large city or learning the status of a parcel in a postal service. Similarly, we observe an unprecedented development in sensing technology, sensors and sensor networks. A variety of sensor types are now available on the market in many domains, from tasting the quality of wine/tea/coffee to determining the temperature, the humidity and the mineral and water content of the ground for agricultural purposes, sensing/monitoring the physiological conditions of drivers/patients etc.. Sensors operate at various frequency bands and locations, e.g., indoor/outdoor, airborne, space-borne, terrestrial, underwater and underground.

Traditional approaches may not be optimal for the integrated provision of these services, because of the allocation of different frequency bands, waveforms and hence different receiver platforms for these services. For example, navigation signals generally contain information about the platform

Convergence of Communications, Navigation, Sensing and Services, 1–30.

identity/location and the time. In addition, sufficiently large bandwidths are allocated to navigation services so as to allow accurate position determination and transmission/reception of navigation signals. Similar arguments may be repeated for communication and sensor systems. In view of this and the fact that modern telecommunication systems support very high data rates, strongly needed integration of digital COmmunications, NAvigation, SENsing and related Services (CONASENSE) looks feasible (see Figure 1.1). The technology is believed to be available for the integration of CONASENSE-related services under realistic scenarios. The emphasis of the CONASENSE Initiative [1], [2] is on the improvement of the quality of life (QoL) of human beings in harmony with the environment using the available and enabling technologies. The CONASENSE mission also includes helping the development of innovative technologies for mid- to far-terms. The QoL is believed to be improved when human beings can choose freely between sufficiently many user-friendly CONASENSE services which do not compromise the user-privacy.

Covering a large domain of research, the CONASENSE has a high potential for a variety of applications and service provision to a large spectrum of users. Consequently, the CONASENSE-related studies may be research-oriented, e.g., related to system architecture, performance evaluation, protocol design, physical layer techniques etc.; application-oriented, e.g., proof-of-concept studies, system or prototype development; and/or service-oriented, including approaches for the provision of a multitude of services. These services may be related to, for example, health, monitoring and protection of the environment, traffic-control and security.

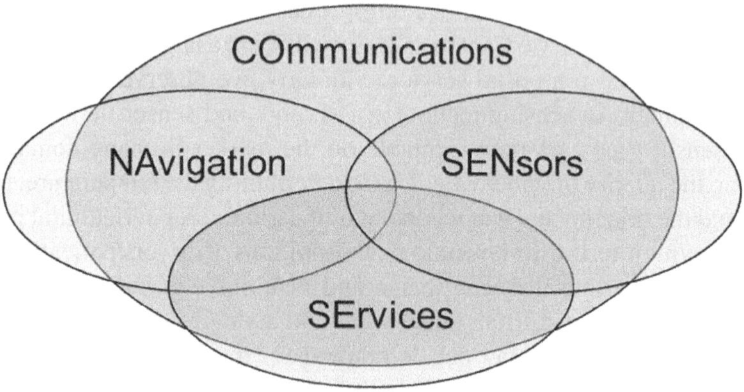

Figure 1.1 The CONASENSE framework

Section 2 will present the requirements for present and future CONASENSE services. Section 3 will provide architectural considerations related to the CONASENSE-related technologies and services with emphasis on areas including e-Health, security, traffic control, environment monitoring and protection. The CONASENSE architecture aims to specify how the infrastructures related to communications, sensing and navigation should interact so as to provide the desired QoL services. Architectural considerations will be presented for the evolution of CONASENSE services with time even though long-term predictions may not be easy. The problems and drawbacks of the current infrastructures and architectures will be reviewed, and backward compatibility issues will be briefly discussed.

1.2 Requirements

QoL improvement by developing novel CONASENSE systems requires strategy for and design of integrated architecture that meets the technical requirements at system level and for all interfaces (terminals, equipment, person to machine, M2M [3–5]) as well as technical requirements for receiver and system design. Hence, integrated synergistic use of CONASENSE-capabilities is mandatory. Flexible, intelligent, and heterogeneous architecture of the system must operate with any kind of enabled devices (i.e., fixed, portable and handheld ones) and control centers, and the services must be provided by, amongst others, satellite and terrestrial segments. The use of advanced cognitive, cooperative and context– and location-aware technologies and distributed system intelligence and application programmable interfaces (API) are strongly desired. The architecture should be intelligent with sensing, learning, decision and action functions, with the exploitation of the benefits of clouds of users; reconfigurable, adaptive, energy efficient; high efficiency, capacity satisfying QoS requirements; support security demands; incorporate context awareness; integrate available technologies and provide the optimization of the architecture for various applications. The middleware should integrate all heterogeneous components (localization, communication and sensing) in a common stratum to implement the basic functionalities for all the services. On top of the middleware the applications should provide the complex system intelligence functionalities.

QoL improvement can be fastened via multi-disciplinary academic and applied research, which requires international institutional cooperation and the presence of test laboratories with state-of-the-art test and measurement equipment. Main criterion for developing novel CONASENSE system demonstrators is that major user organizations indicate the utilization potentials for solving problems in Society and in QoL in particular. Those system developments should be selected which require latest (and new) insights in basic and applied technology and which bring clear visions on "selling attributes", like needs for introducing the system in applications, potential market expectations etc. Feedback on CONASENSE initiatives need to be solicited from user organisations, industry and governmental innovation bodies so as to facilitate fund raising.

Below is presented a brief description of the requirements about the CONASENSE-related services from different perspectives.

1.2.1 Requirements for Terminals/Platform

Users, transceivers, and platforms to be navigated, and targets/parameters to be sensed in-situ or remotely may be listed as space-borne (satellites, unattended aerial vehicles (UAVs), high-altitude platforms), sea-borne, ground-based, underground (earthquake, mines, tunnels etc.), underwater (communications, vehicle navigation, etc.), isolated places (mountains, forests, seas, oceans, deserts), indoor/outdoor, live tissues etc. For example, space-borne telescopes are already being used successfully in radio astronomy. Near-earth orbiting satellites are used in many areas such as remote-sensing, harvest prediction, surveillance and environmental protection. Sensing and telecommunications may also be used for safety purposes, e.g., monitoring disaster and underground mines and tunnels. Similarly, underwater communications, sensing and navigation may have numerous application areas, including navigation of underwater platforms, remote sensing wild life in the seas/oceans/deserts, fishing industry, mineral exploration in the oceans and monitoring and protection of the environment. Similarly, positioning and navigation applications are rapidly increasing in many areas of medicine. The CONASENSE architecture should fulfill the above-listed requirements.

1.2.2 User Requirements

The requirements for CONASENSE-related services are mainly related to the user mobility and the environment. The users would prefer to have low-cost, light-weight, user-friendly, and low power/energy consuming receiving terminals, operating at all/most frequency bands allocated to the CONASENSE-related services with optimized coverage. Availability of these services may impose stringent requirements, e.g., the availability of navigation services might be of vital importance in certain scenarios. The privacy and survivability may be required for many services. The cognition, self-organization capability and adaptability of the terminals against different environmental and operational conditions are strongly desired. In addition to all these, these services must be ubiquitous, reliable and affordable.

1.2.3 Technical Requirements for Receiver and System Design

The frequency bands allocated to CONASENSE services differ depending on the particular application. Consequently, the propagation channels behave differently and the requirements for transceiver design are not the same. Depending on the applications and the related services, the operating frequency bands may cover radio frequency (RF), optical, infrared (IR), acoustics etc. A common (integrated) and interoperable receiver for CONASENSE services is required to operate at different frequency bands, transmit power levels, receiver sensitivities, antenna structures, single- and multi-carrier transmissions techniques and modulation schemes [6–9]. Improved positioning accuracy by integration of the navigation data collected from different sources, such as global positioning system (GPS), Galileo, GLONASS, Wi-Fi, gyroscopes, accelerometers etc., is strongly desired.

Software-defined radio (SDR) may be considered as a strong candidate for an integrated receiver. Present technologies allow the design of agile front-ends with frequency synthesizers of fast hopping and settling times and with low phase noise. Followed by fast analog-to-digital and digital-to-analog convertors, the present signal processing technology facilitates the design of common low-noise and sensitive SDR receivers, by using a DSP chip, ASIC or FPGA, operating at numerous frequency bands and interoperable with different systems [10], [11]. A multi-band SDR for spaceborne communications, navigation, radio science and sensors is reported to support communication, command and telemetry links, high-rate scientific data return links and two-way Doppler navigation. Modularity within core hardware and firmware platforms allow for additional software and software upgradeable features, technology enhancements and implementations with minimal non-recurring engineering costs [12], [13].

In view of the above, system architecture design which would enable integrated CONASENSE services is a challenging task. One of the first issues to be resolved in this context concerns the interoperability with LTE/4G, professional mobile radio (PMR) and/or terrestrial and satellite-based navigation systems [13]. In view of the anticipated diverse applications, direct communications and multi-hop relaying between different user equipment without using base stations may also be required. Other issues to be concerned include broadcasting, multicasting, security, routing, radio resource management, efficient power/energy consumption, network establishment etc.

1.2.4 Energy Requirements

Energy is a valuable resource in CONASENSE-related applications. Design of battery- or electricity-operated wireless communications and sensor nodes are based on continuous flow of power from the energy/power source to the electronic equipment. If electricity or battery is not available on-site, the systems may be operated by harvesting energy from their environment. Energy harvesting implies the collection of energy from ambient sources and converting into electrical energy. However, irrespective of whether the required electric energy is provided by the mains, the battery or energy harvesting, minimization of the consumed energy is strongly desired because of reasons such as cost, equipment life-time, electromagnetic compatibility and for future innovative applications [14].

From an energy perspective, a communication or a sensor node may be considered to be composed of supply and demand sides. A sensor node differs from a communication node mainly by the presence of a sensor. The *demand side* consists of energy consuming units such as a sensor, a signal processing unit, a wireless transceiver and a buffer, either to store the sensed data or the data to be transmitted/received [15], [16]. Transceivers typically use Bluetooth or Zigbee protocols to communicate within a range of maximum 30 m and require output power levels in the order of 2–100 mW. Hence, power levels needed by a sensor node may be in the order of few 100 mWs including all components. The *supply side* of a node consists of energy storage and energy harvester in energy harvesting systems. Using the harvested energy and/or the battery, it communicates via its transceiver with outside world; it receives orders from and transmits the sensed data to its base station. The lifetime of energy harvesting systems is theoretically infinite.

Wireless sensor nodes are usually battery-operated because of the difficulty and/or inconvenience of reaching sensor nodes in remote locations, high cost of maintenance and replacement. Hence, the energy efficiency determines the life-time of battery-operated sensor nodes, which are required to provide independent, sustainable and continuous operation. Battery-operated nodes do not have an energy harvester and the node life-time is limited by the battery capacity. Recently remarkable improvements are observed in power density (W/kg), efficiency, amount of supplied power and the capacity (Amp.-hour) in the area of energy storage [17]. Nevertheless, operation by batteries still has its limitations and may not be suitable for certain applications.

Hence, there is strong demand for energy harvesting systems that can generate their own energy from their environment. Since energy harvesting may not always be available and predictable, energy harvesting systems employ batteries for storing the harvested energy for present/future use. In this respect, the harvesting efficiency and the availability of energy source are the fundamental issues to be considered. Since the present designs are presently based on the continuous flow and ever-presence of the electric energy, these nodes may not operate optimally with energy harvesting and a novel approaches are required for designs with energy harvesting. Therefore, energy harvesting nodes should be designed so as to account for the limitations due to scarcity, non-uniform flow and limited-availability of power in some time intervals which cannot be predicted beforehand. The CONASENSE architecture should allow for energy harvesting in mid- to far-terms.

1.2.4.1 Intelligent Designs for Self-Powered (Energy Harvesting) Nodes

Classical design of sensor/communication nodes is based on the availability of a continuous flow of a constant power level (infinite energy) to the demand side. On one hand, dramatic reductions are strongly desired in power/energy levels dissipated on the demand side of existing wireless nodes. The energy consumption in transceivers may be decreased by reducing the data to be transmitted using source coding, choosing adaptive channel coding and modulation strategies, using efficient transmission scheduling, exploiting power saving modes (sleep/listen) and using energy efficient routing and medium access control [18], [19]. On the other hand, energy harvesting technology is presently far from satisfying present needs. Energy sources may be (un)controllable and/or (un)predictable for energy harvesting; solar energy is predictable but uncontrollable, while RF energy harvesting in a RFID system may be controllable and predictable at the same time. Therefore, limited power/energy that can be harvested sets a constraint on the average power or energy consumed by the demand side for self-powered operation. This implies that energy harvesting, storing and processing technologies should be improved so as to help sustainable and continuous operation.

Even if infinite energy becomes available to the supply side, energy generation may not be continuous and/or rate of generation may be limited. Storing the harvested energy may partially alleviate this problem since it may regulate the power flow. Nevertheless, electronic devices with classical design cannot reliably operate under these conditions. Therefore, energy generation profile of the supply side should be matched to the energy consumption profile of the demand side. This requires a system-level approach involving variation-tolerant architectures, ultra-low voltage and highly digital RF circuits. In addition, one needs DSP architecture and circuits which are energy-efficient, energy-scalable, and robust to variations in the output of the supply side. Energy-scalable hardware may call for techniques for approximate processing, which implies a trade-off between power and arithmetic precision [20]. In wireless sensor networks, the demand side may be designed with sleep/awake periods in synchronism with energy harvesting by the supply side. Energy consumption policy may be optimized in seeking a tradeoff between the throughput and the lifetime of the sensor node [21]. Such approaches are believed to result in more than an order of magnitude energy reduction compared to present systems [22].

In some projects like Pico Radio (Berkeley), μAMPS (MIT), WSSN (ICT Vienna) and GAP4S (UT Dallas), densely populated low-cost sensor nodes are foreseen to operate with power levels of approximately 100 μW; such power levels is believed to be within the capabilities of energy harvesting. Even though dramatic improvements are still needed, rapidly-evolving energy harvesting technologies are believed to be promising for self-powered operation. The CONASENSE architecture in the mid- to long-term should address the energy harvesting problem especially in the mobile communication platforms and wireless sensors.

The use of nanogenerators is foreseen to be used for a variety of applications including intra-body drug delivery, health monitoring, medical imaging, environmental research (air pollution control), military applications (surveillance networks against nuclear, biological, and chemical attacks at nanoscale, and home security), and very high data rate communications. Energy harvesting may enable the widespread use of the nanotechnology [23].

In addition to energy harvesting, the energy problem may be alleviated by using cognitive approaches and distributed and cooperative data processing among communication and sensor nodes. Beamforning in WSN also needs to be considered for CONASENSE architectures with reduced energy consumption.

1.3 Architecture

The CONASENSE architecture is required to satisfy the requirements listed above. In this context, a flexible and modular platform integrating CONASENSE-related services should be able to address a wide range of QoL applications. Such a complex platform operating in different frequency bands/channels and for numerous applications should be user-friendly, energy-efficient and able to operate at different transmit/receive power levels using adaptive modulation/coding and transmission techniques. MIMO techniques, wavelets and ultra-wideband may be considered to render the CONASENSE architecture more flexible. User-friendliness is an important issue for improving the QoL of users of all ages with different needs and education levels. Priority of services and users are thought to be essential in complex platforms.

The security is important not only for navigation, location finding and positioning, but also for sensing and telecommunications. Therefore, the CONASENSE architecture design should also satisfy mid- and far-term security requirements in a user-friendly way without compromising the system performance. Similarly, reliability in normal operations and safety in case of abnormal events and emergency situations should be provided by the architecture.

On the other hand, in present day as well as in the mid- and far-terms, the user privacy will definitely be an essential requirement for the QoL. Some aspects of sensing (medical, biological) may require higher degrees of privacy compared to others. The privacy issue should be tailored and controlled by the user, since it can change depending on user needs and specific applications/situations; hence architecture designs providing flexible privacy degree controlled by the user should be sought.

The architecture should be flexible enough so as to allow the introduction of cognition into the system as new technological developments permit it, especially in mid- to far-terms. The cognitive elements inherent in the integrated system architecture should enable the system to be adaptive and leading to optimized decisions in quasi-real-time according to the user type, channel conditions and applications. In this respect, heuristic approaches for quality improvement of generic services may help the architecture to minimize the time for data collection, signal processing and decision making as well as to allow a trade-off between the optimality of the decisions and the required computational complexity. The architecture should be open so as to ease the introduction of new services/applications as much as possible.

Since the anticipated architecture will operate with various communication, navigation and sensor systems at different frequency bands and in different channels, the operation and/or coexistence in the mid-term with IP should be carefully considered. One should as well consider the feasibility of new network protocols for long-term architectural studies. User-centric architecture should account for the evolution of large numbers of CONASENSE services with time. One needs to carefully consider the optimization of the architecture vis-à-vis the CONASENSE services and whether layerless communications lead to an improved architecture at least for the provision of some selected services.

The integration of CONASENSE functionalities by means of heterogeneous and reconfigurable networks is a breakthrough for the growth of distributed cloud computing and social interaction technologies and a big leap towards the provision of a plurality of services and applications, ideally and in perspective a universal library of services seamlessly provided to users. In the sub paragraphs below various services will get attention: positioning, sensing, e-health, security and emergency services, traffic management and control, environment monitoring and protection, smart power grid. Such new services can improve the QoL of users. The services related to e-health and emergency fields are exemplary applications, where flexible, multi-service and cooperative heterogeneous architectures play a fundamental role. The integration of CONASENSE functionalities in a heterogeneous, flexible, cooperative system is far from being presently available. Design, implementation and deployment of this visionary scenario to provide better services to the third-millennium users is indeed one of the most challenging issue for the scientific community.

For economic and technical considerations, a modular CONASENSE system architecture is strongly desired, meaning that each system is composed of a series of sub-systems in hardware and/or software. Comparable to "SDR", the basis for novel CONASENSE systems may follow a "software-defined-CONASENSE" approach. It should lead to set up a set of standards at the interfaces so that integrated system developments, functionality, and performance can first be tested in software. Embodied software-defined functionalities and on-line testing performance may be specified in order to meet the overall system specifications at system and sub-system levels. The tests can be made even before hardware developments start, but also during various phases of the (sub-) system developments. Different technological institutes may take the responsibility for the progress and development for the specifications, acceptance, integration and technical testing of different sub-systems while some others assume the responsibility for performance testing of the integrated system and user interface performance.

Novel architecture design approaches should be jointly considered with backward compatibility issues since new systems need to be integrated with the existing ones. Similarly, horizontal integration is also needed between networks for communications, navigation and sensors. Potential architecture solutions should also be based on trade-offs between the cost and innovative content.

Efforts for modeling the architectures facilitate the comparison of both existing and proposed architectures in order to determine sources of technical challenges in implementation and facilitate the estimation of the implementation and operational costs. Risk assessment evaluations are also necessary. Developing a figure-of-merit help enabling the quantitative evaluation of each architecture and operational process option. Modular system designs, approches for system operation as well as software and hardware fixes and maintenance should be should be carefully considered [24].

1.3.1 Positioning

Positioning, location finding and navigation plays a crucial role in CONASENSE applications. Accurate, reliable and real-time positioning is a serious issue in the operation of location-aware services, e.g., in the formation and self-organization of ad hoc networks [25], in navigation and sensing [26], emergency conditions etc.

Existing positioning systems have different waveforms, operational frequencies and capabilities. In addition, different frequency bands are allocated to navigation, communications and sensing systems. Therefore, the CONASENSE architecture should optimize data collection and decision making in central or distributed ways for improved positioning accuracy. The recent advances in positioning techniques is believed to improve the positioning accuracy in indoor and outdoor environments and pave the way for many innovative CONASENSE applications even in the near future. Recent technological developments e.g. in micromechanical systems (MEMS), enables the development of gyroscopes and accelerometers at smaller sizes to be incorporated in mobile terminals. Similarly, the use of multiple-input multiple-output (MIMO) techniques is expected to improve not only the performance of the communication systems but also help for accurate position and time estimation. Intense research efforts are going on for the integrated design of navigation systems for improved time and position estimation. Telecommunications, sensing and navigation communities are interested in bio-inspired algorithms [27] for improving the performance of CONASENSE-related services. Evolution-perfected bio-algorithms for colony life, migration of fishes, bees [28], ants, birds and herds inspire the scientists to exploit bio-inspired algorithms more aggressively.

Positioning systems are usually categorized as network-based or mobile-based depending on the location where position calculations are performed. Calculations for position estimation may be mobile-based, when the positioning information is extracted from the received signals, or network-based if information collected through reference terminals is processed at a central unit. Positioning systems may be either terrestrial-based and used for both outdoor and indoor environments, or satellite-based, which offer global coverage but generally serve to only outdoor users.

Satellites can play a leading role in CONASENSE services due to their immunity against ground-based catastrophic events and for their ability of collecting information created by sensors deployed on the surface of the earth and the sky, if necessary. They can provide communications, sensing and enable assisted localization, combining information from positioning satellites and terrestrial terminals. Efficient architecture design of hybrid terrestrial-satellite positioning systems and their integration with communication systems is a challenging problem. Multiband receiver antennas are needed for operation in the frequency bands allocated to navigation, sensing and communication systems. Transmit antennas should be designed so as to produce signals with isotropic power spectral density within global coverage for navigation receivers [29].

Indoor positioning, integration of positioning with payment systems, positioning in live tissues, e.g., in human body, underwater positioning, positioning in tunnels, positioning of chemical pollutants in the air are among the numerous areas to be discovered.

1.3.2 Sensing

Recent advances in digital technology enabled the development and production of high resolution, low-power, environment-friendly, long-life, low-cost and small-size sensors [30], [31]. Consequently, we observe in our everyday life various sensor types, including RFID-, MEMS-, biometric-, acoustic-, video-sensors and so on. Sensors are used in a very large spectrum of applications, including health monitoring [32], [33], underwater acoustic networks [34], smart grid applications [35], agriculture [36], emergency applications, automotive industry [37] etc. Therefore, applications related to sensing will definitely have an increasing importance in CONASENSE-related applications.

Sensors may be used for sensing locally or remotely; the information collected by the sensed signals may be processed in situ, in a distributed fashion, or at a fusion center [38], [39]. Multiple sensors may be employed for cooperative sensing when the data collected by a single sensor does not meet the requirements. Data collection, processing and management architecture and techniques, e.g., sensor fusion, data fusion and/or information fusion, and making an optimum multi-criteria decision concerning the sensed data, need to be carefully considered in the design of the CONASENSE architecture.

Heterogeneous networks of sensors may be remotely located from each other and operate at different frequency bands. For example, in a maritime environment, there are various technologies for detection and locating objects such as coastal radar, sonar, video camera, IR, automatic identification system, automatic vessel locating. In addition, these signals should be processed with HF communications, intelligence data etc. Moreover, centralized or decentralized fusion of the information provided by various sensors may be required to ensure reliable performance for handling complex scenarios due to temporary loss of availability, error, limits of coverage etc. A multi-sensor tracking and information fusion methodology /architecture is needed to harness the effectiveness of multiple sensor information [40].

Recent research efforts on electronic nose, electronic eye and electronic tongue lead to versatile and innovative applications. Sensing and monitoring volatile organic compounds, atmospheric pollution, hazardous gases, chemicals and explosives may be cited among the applications for security and environmental protection. In health-related applications, one may cite the diagnosis of lung cancer at early stage, identification of urinary tract infection and helicobacter as well as detection, discrimination and monitoring of drug, drug users and smokers. Sensor systems are also used for building artificial nose, tongue and eye for robots and other applications as cited in the literature [41], [42].

A remarkable area that has to be emphasized in the future CONASENSE applications is the amount of data exchanged in the wireless sensor networks (WSN). The data across all levels of the network may be generated by smart sensing systems, supervisory control and data acquisition systems, wide area monitoring systems, and other sensing/monitoring devices. The huge amounts of data and information need to circulate and be stored among control centers, devices and users. Therefore, the use of data compression techniques will be desirable to help mitigating the burden of the communication among CONASENSE sensors and control systems. To this end, the information acquired by the sensors should be compressed at the sending terminals as much as possible, before sending through the wireless communication system. The compression should preserve the valuable information contained in the data, and the compressed data- when received at receiving terminals- should be perfectly reconstructed too for analysis. In this regard, the Wavelet technology for data compression is of paramount importance [43] for the future CONASENSE applications. With this technology data can be compressed before it is sent out in order to mitigate the data congestion in the intelligent sensing network. Due to the nature of wavelets, the technique is beneficial not only in reducing white noise but also in the mitigation of a wide range of interferences which are present in different application scenarios of CONASENSE.

Another important issue in sensing area, in general, and in the wireless sensor networks, in particular, is the self-configuration and self-organization feature of these networks, or, in other words, the cognitive aspect of smart WSN. Intelligent CONASENSE sensors are aimed to transform the already existing network into an advanced, cognitive and decentralized infrastructure. The cognitive characteristic will be a viable choice for future WSN. In this context the cognitive radio communication technique is strongly needed for the implementation of smart WSN on the physical system level dealing with information and communication technologies (ICT) hardware and technical interoperability. The ICT ideas of cognition and intelligence are required to make WSN smart, and to ensure their stability, reliability, and security. The cooperative and self-organization aspect of CONASENSE is the salient aspect of WSN of future that has to be deeply researched.

Another important research area in the smart sensing is its *greenness*. As the smart WSN of the future will be sustainable no need to say that the automation process and the sensors communication and control in these intelligent networks should be green as well. Therefore, there will be an emergent need for developing energy efficient and green sensor technologies that optimize power consumption even while guaranteeing a desirable quality of service and a robust and secure communication/control performance. Green ICT technologies will certainly be on the agenda for future research and development of CONASENSE systems and networks.

Currently compressed sensing is a rapidly emerging field of research. Establishment of WSN in the light of mathematical theory of compressed sensing is the state-of-the art. Due to sparsity of the sensed signals, proper bases are used for compressed sensing while satisfying the criteria for compressed sensing. The promising achievements in this area are the reduction of number of sensing elements and measurements, the reduction of the complexity of the sensing method, optimized use of sensing power, and the optimized number of sensing nodes [43].

1.3.3 e-Health

In view of rising costs in healthcare, increasing percentage of ageing population and the fact that many patients needing health-monitoring do not necessarily require hospitalisation, one needs to look for new approaches for the provision of low-cost (preventive) health services, especially for disabled, elderly and chronically ill patients [44–45]. On the other hand, the progress in ICT, biotechnologies and nanotechnologies accelerate innovations in the field, and lead to miniaturization and large-scale production of efficient and affordable products. Standardization that will ensure interoperability between devices and information systems will open up the way for large-scale and cost-effective deployment of e-health systems [46]. The CONASENSE may therefore play a major role in the provision of e-health services, hence for the QoL improvement.

Thanks to recent advances in sensor technology and networks, the human health can be monitored by collecting data on specific physiological indicators (e.g. blood glucose level, blood pressure, electrocardiogram and electroencephalogram, portable magnetic resonance images, implantable hearing aid etc.), via in-, on-, and/or out-body sensors. An electronic system is reported in [47] that achieve thicknesses, effective elastic moduli, bending stiffnesses, and areal mass densities matched to the epidermis. Unlike traditional wafer-based technologies, laminating such devices onto the skin leads to conformal contact and adequate adhesion based on van der Waals interactions alone, in a manner that is mechanically invisible to the user. The system incorporates electrophysiological, temperature, and strain sensors, as well as transistors, light-emitting diodes, photodetectors, radio frequency inductors, capacitors, oscillators, and rectifying diodes. Solar cells and wireless coils provide options for power supply. This technology is designed and manufactured to measure electrical activity produced by the heart, brain, and skeletal muscles.

Such systems typically perform sensing, data collection with user profile information, including data aggregation, data visualization, and analysis/alerting functions for the health professionals. The data, which is usually collected at a hub, is periodically transmitted to a server through a gateway (using IP). The database in the server may be used for preventative health care, physiological/functional monitoring, chronic disease management, and assessment of the QoL, e.g., fitness, diet or nutrition monitoring applications. In view of the multiplicity and mobility of sensors and users, association of the health monitoring data with patients requires serious consideration [48].

In e-health systems, the patients or sensing systems may update their data in real time through Internet. Hence, the record of a patient becomes available to authorized professionals anytime anywhere, for real-time monitoring and intervention in emergency. Such systems also provide new forms of interaction and coordination between health professionals and lead to novel scientific approaches for medical applications.

E-health systems provide mobility to the patients via mobile health-monitoring devices [49]. Patient mobility also calls for wearable, outdoor and home-based applications. Physical and health conditions of patients can be monitored in real-time using sensors observing the environment and those that measure physiological parameters of the patient at home and hospital environments. Similar approaches may be followed for the safety of workers. Indoor positioning and tracking systems may be used in hospitals, for example to track expensive equipment, and to guide patients and health professionals inside the hospitals for more efficient and timely services.

The health data can be stored on the sensor nodes and analyzed offline, while emergency situations and reports may be made available to health workers through a remote database. Hence, stored data from multiple patients may be utilized for geographic and demographic analyses [50]. Mobility solutions for wireless body area networks (BANs) for healthcare are already available, through communications over low-power Personal Area Networks (6LoWPANs) using Internet protocol Ipv6 [51]. These systems can provide preventive healthcare, enhanced patient–doctor interaction and information exchange. Continuous health-monitoring allows immediate intervention in case of an emergency. Positioning, tracking and monitoring patients with asthma, diabetics, heart disease, Alzheimer disease, obesity [52] as well as visually impaired people, ambulance systems, small children, robotic wheelchairs, pregnancy, blood pressure, artificial arms/legs, drug addiction are believed to be important issues.

Modelling the physical channel in body-area networks (BANs) is the topic of intense research efforts [53]. Optimal use of relay nodes, adaptive approaches for managing outages and retransmissions, cross-layer optimization to share information between physical and media-access control (MAC) layers will definitely improve the overall system performance. The MAC-layer operation in BANs for e-health is addressed in the IEEE 802.15.6 draft standard for BAN [54]. For example, intra-body communication for continuous-monitoring of patients with artificial heart embodies serious issues, e.g., the operation of an antenna and the propagation of electromagnetic waves in human body, which is a nonhomogeneous lossy medium. Similarly, reliable signal transmission between sensors on the body under shadowing is still among the areas of interest. Positioning with high precision in the human body, which is strongly desired for surgical operations, is believed to be possible in the near/mid-term.

The physiological data collected in e-health systems is bidirectional and distributed through Internet and/or in heterogeneous networks to all interested parties. The CONASENSE architecture should address the problems related to protocol design, accuracy, reliability, data security, protection, privacy of diagnoses, range, operation time, and interoperability between medical devices [55–57]. These studies should be supported by databases, intelligent decision support algorithms and programming languages. Recent efforts on medical device interoperability resulted in a standard ISO/IEEE 11073 PHD [58] for communications between health devices such as USB, Bluetooth, and ZigBee.

CONASENSE architecture should also provide intelligent solutions for wireless e-health applications including healthcare telemetry and telemedicine. Even the patients at some remote locations and/or unable to reach a nearby health center may be monitored and managed; remote diagnosis and emergency intervention can be accomplished by tele-medicine. Improved and low-cost healthcare services may be provided to poor and geographically remote patients by exploiting new technologies. In that context, such services may provide a low-cost healthcare solution in less developed geographic regions in the world.

Terrestrial and satellite segments of the CONASENSE architecture provide the deployment of interconnected integrated and interoperable telecommunications network for e-health applications. e-Health services are provided by an Interactive Service Platform, including real-time audio and video interactions among patients, specialists and health service providers. Both citizens and physicians can access the interactive Service Platform from different locations (e.g. Health Points, Hospitals, Home) regardless of the chosen access technology, either satellite or terrestrial. The Service Platform will share health information among different applications and services (Self-Care and Assisted). It is based on the Health Integration Engine (HIE): this middleware guarantees the information exchange between the e-Health subsystems Personal Health Media (PHM), Electronic Health Record (EHR), Electronic Clinical Research Form (ECRF), Clinical Health record (CHR) and the user access points.

1.3.4 Security and Emergency Services

Professional mobile radio (PMR) systems, e.g., APCO and TETRA, are employed by police, fire departments, ambulance systems etc. for security and emergency applications. Compared with ubiquitous commercial mobile radio systems, these systems have some additional requirements for survivability in disaster scenarios, operation in relay mode without needing base stations, larger coverage areas (higher transmit powers) etc. Interoperability between mobile radio and PMR systems is strongly desired. 4G systems may also provide services to the PMR users via virtual private networks. Features like relaying and survivability may be provided through diversity and coordination between base stations. Incorporation of sensing capability and accurate time- and positioning estimation in mobile radio systems may put them in a very strong position for monitoring, and management of disaster, emergency, mine-accidents, earthquake, fire-fighting, police patrolling, and intruder/fraud detection. Rapid and accurate position estimation/navigation is very often needed in military applications.

With accurate position estimation, a variety of applications and services, such as location sensitive billing, and improved traffic management for cellular networks may become feasible. Positioning of a mobile terminal is considered to be critical for position-aware services such as such as E-911 in USA and E-112 in European Union (EU) for emergency calls [59], [60]. Noting that mobile-originated emergency calls are continually increasing and about 50% of all emergency calls in the EU are originated by mobile users, location estimation of a mobile user making an emergency call is strongly desired.

1.3.5 Traffic Management and Control

CONASENSE may also have a significant contribution in the domain of traffic management and control. For example, monitoring and management of highway/tunnel/bridge traffic which may need to be diverted, under congestion, to alternative itineraries may save valuable money and time and reduce pollution. Intelligent transport systems (ITS) will significantly alleviate the urban traffic via inter-vehicular communications, communications between the terminals along the roads, and broadcasting to vehicles the last-minute traffic information. Controlling the distance between vehicles on the road by onboard radars under rain/snow/fog is strongly desired [61–66]. Similarly monitoring and navigation of the traffic in railways, harbors/ports/seas, e.g., yachts, ships etc., air traffic (air traffic control and taxi) and UAV's are among the application areas of the CONASENSE. Monitoring and controlling the border traffic between countries may also be used to prevent illegal border crossings. Traffic control in shopping malls, banks, concert halls etc. may be desirable for statistical purposes as well as for security reasons. Concepts for traffic control, e.g., monitoring the migration paths, times and the density of wild animals, could be valuable for the protection of wild life. Similar arguments may be repeated for the monitoring of the farm animals. In summary, CONASENSE architecture should provide intelligent and flexible solutions in the area of traffic management and control.

1.3.6 Environment Monitoring and Protection

Rapid growth of the world population, high cost of transforming already established and highly polluting manufacturing plants to become more environment-friendly constitutes serious threats to our planet. Fortunately, recent advances in CONASENSE-related technologies enable us to follow more environment-friendly approaches at lower costs. Higher resolutions in positioning and remote sensing is promising for monitoring and assessment of earth resources, agricultural harvest, forests, seas, wild life, water resources, weather/climate, ozone layer, electromagnetic and chemical pollution. The CONASENSE architecture should address this problem, which is believed to be increasing importance in mid- to far-terms, by integration of air-borne and ground-based positioning and remote sensing platforms operating in various frequency bands.

1.3.7 Smart Power Grid

Smart power grid modernizes the current electricity delivery system by integrating ICT into generation, delivery, control and consumption of electrical energy for enhanced robustness against failures, efficiency, flexibility, adaptability, reliability and cost-effectiveness. In that sense, smart grid embodies a fusion of different technologies where electrical power engineering meets sensing, ICT, positioning, control etc. [67]. The communication protocols to be deployed for this purpose should modernize and optimize the operation of the power grid and provide an accessible and secure connection to all network users, especially for efficient use of distributed energy sources [68]. Through communications between interconnected sensor nodes, the smart grid controls equipment and energy distribution, isolates and restores power outages, facilitates the integration of renewable energy sources into the grid and allows users to optimize their energy consumption.

Design of end-to-end QoS resource control architectures [69] and general cyber-physical systems [61], efficient schemes for admission control, monitoring/control of the smart grid and the fluctuations of the power load is believed to be of prime importance for smart power grid networks [70]. Reliability and security of integrated communication network of the Smart Grid should be guaranteed in very adverse power line channels, which suffer high attenuation, multipath, impulse noise, harmonics and distortion [71], and smart attacks [72]. In addition to more reliable power line channel models [60], robust modulation, coding, encryption [67], [72] and transmission techniques and computational models are needed for optimized physical layer performance.

1.4 Conclusions

Emergence of enabling technologies and rapidly increasing demands for services related to communications, sensing and navigation provides a good opportunity for the integration of existing CONASENSE-related systems and to design a novel flexible architecture so as to meet present and future requirements. To achieve this goal, it is critical to identify the requirements for energy, terminal/platform and receiver/system design concerning diverse application areas including e-health, security/emergency services, traffic management and control, environmental monitoring and protection, and smart power grid. Minimization of energy consumption and energy harvesting deserve special attention in the novel CONASENSE architecture design mainly because of requirements for mobility, high-data rate communications and signal processing and green communications. The novel CONASENSE architecture design should address problems and drawbacks of the existing infrastructures/architectures and be sufficiently flexible for future/potential developments. Consequently, the design of the CONASENSE architecture should be carried out so as not only to integrate existing and novel communications, navigation and sensing services but also to provide smooth transition between existing and new systems in hardware and software.

Acknowledgment

The authors would like to thank the **CONASENSE WG2 Members** Ernestina Cianca, University of Rome, Prof. Ramjee Prasad, Aalborg University, Prof. Enrico Del Re, University of Firenze, Prof. Vladimir Poulkov, University of Sofia, Prof. K.C. Chen, National Taiwan University, Dr. Nicola Laurenti, University of Padua, and Dr. Silvester Heijnen, Christian Huygens Laboratories, for the fruitful discussions and their presentations during the WG2 meetings which inspired the authors.

References

[1] www.conasense.org

[2] M. Safak, Potential Applications and Research Opportunities in the CONASENSE Initiative, Chapter 5 in CONASENSE Communications, Navigation, Sensing and Services, River Publishers, Alborg, 2013.

[3] S.Y. Lien, S.-M. Cheng, S.-Y. Shih and K.-C. Chen, "Radio Resource Management for QoS Guarantees in Cyber-Physical Systems," Special Issue on Cyber-Physical Systems, IEEE Transactions on Parallel and Distributed Systems, vol. 23, no. 9, pp. 1752–1761, September 2012.

[4] P.Y. Chen, S.M. Cheng, K.C. Chen, "Smart Attacks in Smart Grid Communication Networks", IEEE Communications Magazine, vol. 50, no. 8, pp. 24–29, August 2012.

[5] K.C. Chen, S.Y. Lien, "Machine-to-machine communications: Technologies and challenges", to appear in the Ad Hoc Networks, 2013.

[6] Bagheri, R., et al., "Software-Defined Radio Receiver: Dream to Reality," IEEE Communications Magazine, Vol. 44, No. 8, August 2006, pp.111–118.

[7] Valls, J., T. Sansaloni, A. Pérez-Pascual, V. Torres and V. Almenar, "The Use of CORDIC in Software Defined Radios: A Tutorial," IEEE Communications Magazine, Vol.44, No. 9, September 2006, pp.46–50.

[8] Minde, G. J., et al., "An Agile Radio for Wireless Innovation," IEEE Communications Magazine, Vol.45, No.5, May 2007, pp.113–121.

[9] Björkqvist, J., and S. Virtanen, "Convergence of Hardware and Software in Platforms for Radio Technologies," IEEE Communications Magazine, Vol. 44, No. 11, November 2006, pp.52–57.

[10] Alluri, V. B., J. R. Heath, and M. Lhamon, "A New Multichannel, Coherent Amplitude Modulated, Time-Division Multiplexed, Software-Defined Radio Receiver Architecture, and Field-Programmable-Gate-Array Technology Implementation," IEEE Trans. Signal Processing, Vol. 58, No. 10, 2010, pp. 5369–5384.

[11] Giannini V., et al., "A 2-mm 0.1–5 GHz Software-Defined Radio Receiver in 45-nm Digital CMOS," IEEE Journal of Solid-State Circuits, Vol. 44, No. 12, December 2009, pp. 3486 – 3498.

[12] M. Lucente et al., PLATON: Satellite remote sensing and telecommunication by using millimeter waves, 2012 IEEE ESTEL Conference.

[13] Haskins, C. B., and W. P. Millard, "Multi-band Software Defined Radio for Space-born Communications, Navigation, Radio Science, and Sensors," IEEE Aerospace Conf., 2010, pp.1–9.

[14] M.Safak, Wireless Sensor and Communication Nodes with Energy Harvesting, CONASENSE Journal, vol. 1, pp.47–66, January 2014, doi: 10.13052/jconasense2246–2120.113.

[15] Niyato, D., E. Hossain, M.M. Rashid and V. K. Bhargava, "Wireless sensor networks with energy harvesting technologies: a game-theoretic approach to optimal energy management," IEEE Wireless Communications, August 2007, pp. 90–96.

[16] Nakajima, N., Short-range wireless network and wearable bio-sensors for healthcare applications, 2nd Int. Symposium on Applied Sciences in Biomedical and Communication Technologies (ISABEL 2009), 2009, pp.1–6

[17] Sudevalayam, S., and P. Kulkarni, "Energy Harvesting Sensor Nodes: Surveys and Implications," IEEE Communications Surveys & Tutorials, vol.13, no.3, 3rd Quarter 2011.

[18] V. Sharma, U. Mukherji, V. Joseph and S. Gupta, "Optimal energy management policies for energy harvesting sensor nodes," IEEE Trans. Wireless Communications, vol. 9, no.4, pp.1326–1336, April 2010.

[19] Joseph, V., V. Sharma, and U. Mukherji, "Optimal sleep-wake policies for an energy harvesting sensor node," IEEE ICC, 2009.

[20] R. Amirtharajah, J. Collier, J. Siebert, B. Zhou, and A. Chandrakasan, "DSPs for energy harvesting sensors, Applications and Architectures," IEEE Pervasive Computing, July-Sept. 2005, pp. 72–79.

[21] Tacca, M., P. Monti and A. Fumagalli, "Cooperative and reliable ARQ protocols for energy harvesting wireless sensor nodes," IEEE Trans. Wireless Communications, vol.6, no.7, pp. 2519–2529, July 2007.

[22] Chandrakasan, A. P., D. C. Daly, J. Kwong and Y. K. Ramadass, "Next-generation micro-power systems," IEEE Symposium on VLSI circuits, Digest of technical papers, 2008, pp.2–5.

[23] Jornet, J. M., and I. F. Akyıldız, "Joint energy harvesting and communication analysis for perpetual wireless nanosensor networks in the Terahertz band," IEEE Trans. Nanotechnology, vol.11, no.3, pp.570–580, May 2012.

[24] J. M. Reinert and P. Barnes, Challenges of integrating NASAs space communication networks, 2013 IEEE International Systems Conference (SysCon), pp.475–482.

[25] Mayorgaet, C. L. F., et al., "Cooperative Positioning Techniques for Mobile Localization in 4G Cellular Networks," IEEE Int. Conference on Pervasive Services, 2007, pp. 39–44.

[26] Gezici, S., "A Survey on Wireless Position Estimation," Wireless Personal Communications, vol. 44, 2008, pp. 263–282.

[27] Stauffer, A., D. Mange and J. Rossier, "Design of Self-organizing Bio-inspired Systems," Second NASA/ESA Conference on Adaptive Hardware and Systems (AHS), 2007, pp. 413 – 419.

[28] Bitam, S., M. Batouche, E.-G. Talbi, "A Survey on Bee Colony algorithms," 2010 IEEE Int. Symposium on Parallel and Distributed Processing, Workshops and PhD Forum (IPDPSW), 2010, pp. 1–8.

[29] Roederer, A. G., "Antennas for Space: Some Recent European Developments and Trends," 18th Int. Conf. Applied Electromagnetics and Communications (ICECom), 2005, pp. 1–8.

[30] Fowler, K., "Sensor Survey Results: Part 1. The Current State of Sensors and sensor Networks," IEEE Instrumentation and Measurement Magazine, Vol.12, No. 1, February 2009, pp. 39–44.

[31] Fowler, K., "Sensor Survey Results: Part 2. Sensors and Sensor Networks in Five Years," IEEE Instrumentation and Measurement Magazine, Vol.12, No. 2, April 2009, pp.40–44.

[32] Wang, C. H., Y. Liu, M. Desmulliez and A. Richardson, "Integrated Sensors for Health Monitoring in Advanced Electronic Systems," 4th Int. Design and Test Workshop (IDT), 2009, pp.1–6.

[33] Fernandez, J. M., J. C. Augusto, R. Seepold and N. M. Madrid, "A Sensor Technology Survey for a Stress-Aware Trading Process," accepted for publication in IEEE Trans. Systems, Man and Cybernetics- Part C: Applications and Reviews, 2011.

[34] Garcia, J. E., "Positioning of Sensors in Underwater Acoustic Networks," Proc. MTS/IEEE Oceans, 2005, vol. 3, pp. 2088 – 2092.

[35] Yang, Y., F. Lambert and D. Divan, "A Survey on Technologies for Implementing Sensor Networks for Power Delivery Systems," IEEE Power Engineering Society General Meeting, 2007, pp. 1–8.

[36] Kalaivani, T., A. Allirani, P. Priya, "A Survey on Zigbee Based Wireless Sensor Networks in Agriculture," 3rd Int. Conf. Trends in Information Sciences and Computing (TISC), 2011, pp. 85–89.

[37] Fleming, W. J., "Overview of Automotive Sensors," IEEE Sensors Journal, Vol.1, No.4, December 2001, pp.296–308.

[38] Nicosevici, T., R. Garcia, M. Carreras and M. Villanueva, "A Review of Sensor Fusion Techniques for Underwater Vehicle Navigation," MTTS/IEEE Techno.- Oceans'04, Vol.3, 2004, pp.1600–1605.

[39] Zhao, X., Q. Luo and B. Han, "Survey on Robot Multi-sensor Information Fusion Technology," 7th World Congress on Intelligent Control and Automation (WCICA'2008), 2008, pp. 5019–5023.

[40] C. Gunasekara, C. Keppetiyagama, N. Kodikara, C. Uduwarage, D. Sandaruwan, K R Senadheera, and J U Gunaseela, Sensor Information Fusion Architecture for Virtual Maritime Environment, The International Conference on Advances in ICT for Emerging Regions (ICTer) 2012, pp. 62–66.

[41] Chang J. B., and V. Subramanian, "Electronic Noses Sniff Success," IEEE Spectrum, Vol. 45, No.3, March 2008, pp. 51–56.

[42] Tang, K.-T., S.-W. Chiu, M.-F. Chang, C.-C. Hsieh and J.-M. Shyu," A Low-Power Electronic Nose Signal-Processing Chip for a Portable Artificial Olfaction System", IEEE Trans. Biomedical Circuits and Systems, Vol. 5, No. 4, 2011, pp. 380–390.

[43] H. Nikookar, Wavelet Radio: Adaptive and Reconfigurable Wireless Systems Based on Wavelets, Cambridge University Press, 2013.

[44] Cova, G., et al., "A Perspective of State-of-the-Art Wireless Technologies for E-health Applications," IEEE Int. Symposium on IT in Medicine & Education (ITIME), Vol. 1, 2009, pp. 76–81.

[45] Aragues, A., et al., "Trends and Challenges of the Emerging Technologies Toward Interoperability and Standardization in E-health Communications," IEEE Communications Magazine, Vol. 49, No.11, November 2011, pp. 182 – 188.

[46] Agoulmine, N., P. Ray, and T.-H. Wu, "Efficient and Cost-Effective Communications in Ubiquitous Healthcare: Wireless Sensors, Devices

and Solutions," (Guest Editorial), IEEE Communications Magazine, Vol.50, No.5, May 2012, pp. 90–91.

[47] D.-H. Kim et al., Epidermal electronics, Science, vo.333, pp.838–843, 12 August 2011.

[48] Chowdhury, M. A., W. Mciver Jr. and J. Light, "Data Association in Remote Health Monitoring Systems," IEEE Communications Magazine, Vol. 50, No. 6, June 2012, pp. 144–149.

[49] Chan, V., P. Ray, and N. Parameswaran, "Mobile E-health Monitoring: An Agent-Based Approach," IET Communications, Vol. 2, No. 2, 2008, pp. 223–230.

[50] Viswanathan, H., B. Chen and D. Pompini, "Research Challenges in Computation, Communication, and Context Awareness for Ubiquitous Healthcare," IEEE Communications Magazine, Vol.50, No.5, May 2012, pp.92–99.

[51] Caldeira, J. M. L. P., J. J. P. C. Rodrigues and P. Lorenz, "Toward Ubiquitous Mobility Solutions for Body Sensor Networks on Healthcare," IEEE Communications Magazine, Vol.50, No.5, May 2012, pp. 108–115.

[52] Mitra, U., et al., "KNOWME: A Case Study in Wireless Body Area Sensor Network Design," IEEE Communications Magazine, Vol.50, No.5, May 2012, pp.116–125.

[53] Ullah, S., et al., "A Comprehensive Survey of Wireless Body Area Networks," J. Med. Syst., Springer, August 2010, pp.1–30.

[54] Boulis, A., D. Smith, D. Miniutti, L. Libman and Y. Tselishchev, "Challenges in Body Area Networks for Healthcare," IEEE Communications Magazine, Vol.50, No.5, May 2012, pp.100–106.

[55] Noimanee, K., et al., "Development of E-health Application for Medical Center in National Broadband Project," Biomedical Engineering International Conference (BMEiCON), 2011, pp. 262–265.

[56] Nita, L., M. Cretu, and A. Hariton, "System for Remote Patient Monitoring and Data Collection with Applicability on E-health Applications," 7th Int. Symposium Advanced Topics in Electrical Engineering (ATEE), 2011, pp.1–4.

[57] Ying, S., and J. Soar, "Integration of VSAT with WiMAX Technology for E-health in Chinese Rural Areas," 2010 Int. Symp. Computer Communication Control and Automation (3CA), Vol. 1, pp. 454–457.

[58] ISO/IEEE11073 — Personal Health Devices Standard (X73PHD), Health Informatics [P11073–00103, tech. rep., overview] [P11073–104zz. Device Specializations] [P11073–20601, Application Profile — Optimized Exchange Protocol], http://standards.ieee.org.

[59] Federal Communications Commission (FCC) Fact Sheet, "FCC Wireless 911 Requirements", 2001.
[60] EU Institutions Press Release, "Commission Pushes for Rapid Deployment of Location Enhanced 112 Emergency Services," DN:IP/03/1122, Brussels, Belgium, July 2003.
[61] Hartenstein, H., and K. P. Laberteaux, "A Tutorial Survey on Vehicular Ad Hoc Networks," IEEE Communications Magazine, Vol. 46, No. 6, June 2008, pp. 164–171.
[62] Karagiannis, G., et al., "Vehicular Networking: A Survey and Tutorial on Requirements, Architectures, Challenges, Standards and Solutions," IEEE Communications Surveys & Tutorials, Vol. 13, No. 4, Fourth Quarter 2011, pp.584–616.
[63] Sichitiu, M. L., and M. Kihl, "Inter-Vehicle Communication Systems: A Survey," IEEE Communications Surveys & Tutorials, Vol. 10, No. 2, 2nd Quarter 2008, pp. 88–105.
[64] Suthaputchakun, C., and Z. Sun, "Routing Protocol in Intervehicle Communication Systems: A Survey," IEEE Communications Magazine, Vol.49, No.12, December 2011, pp. 150–156.
[65] Acosta-Marum, G., and M. A. Ingram, "Six Time- and Frequency-Selective Empirical Channel Models for Vehicular Wireless LANs," IEEE Vehicular Technology Magazine, Vol. 2, No. 4, Dec. 2007, pp.4–11.
[66] Molisch, A. F., F. Tufvesson, J. Karedal, and C.F. Mecklenbrauker, "A Survey on Vehicle-to-Vehicle Propagation Channels," IEEE Wireless Communications, vol.16, no.6, December 2009, pp.12–22 Fadlullah, Z. Md., et al., "Toward Secure Targeted Broadcast in Smart Grid," IEEE Communications Magazine, Vol.50, No.5, May 2012, pp.150–156.
[67] Fadlullah, Z. Md., et al., "Toward Secure Targeted Broadcast in Smart Grid," IEEE Communications Magazine, Vol.50, No.5, May 2012, pp.150–156.
[68] Lloret, J., P. Lorenz and A. Jamalipour, "Communication Protocols and Algorithms for the Smart Grid," IEEE Communications Magazine, Vol.50, No.5, May 2012, pp.126–127.
[69] Vallejo, A., A. Zaballos, J.M. Selga and J. Dalmau, "Next-generation QoS Control Architectures for Distributed Smart Grid Communication Networks," IEEE Communications Magazine, Vol.50, No.5, May 2012, pp.128–134.
[70] Zhou, L., J. J. P. C. Rodrigues and L. M. Oliveira, "QoE-driven Power Scheduling in Smart Grid: Architecture, Strategy, and Methodology," IEEE Communications Magazine, Vol.50, No.5, May 2012, pp. 136–141.

[71] Oksman, V., and J. Zhang, "G.HNEM: The New ITU-T Standard on Narrowband PLC Technology," IEEE Communications Magazine, Vol.49, No.12, December 2011, pp.36–44.
[72] Marmol, F. G., C. Sorge, O. Ugus, and G. M. Perez, "Do Not Snoop My habits: Preserving Privacy in the Smart grid," IEEE Communications Magazine, Vol.50, No.5, May 2012, pp. 166–172.

Biographies

Mehmet Safak received the B.Sc. degree in Electrical Engineering from Middle East Technical University, Ankara, Turkey in 1970 and M.Sc. and Ph.D. degrees from Louvain University, Belgium in 1972 and 1975, respectively.

He joined the Department of Electrical and Electronics Engineering of Hacettepe University, Ankara, Turkey in 1975. He was a postdoctoral research fellow in Eindhoven University of Technology, The Netherlands during the academic year 1975-1976. From 1984 to 1992, he was with the Satellite Communications Division of NATO C3 Agency (formerly SHAPE Technical Centre), The Hague, The Netherlands, as a principal scientist. During this period, he was involved with various aspects of military SATCOM systems and represented NATO C3 Agency in various NATO committees and meetings. In 1993, he joined the Department of Electrical and Electronics Engineering of Eastern Mediterranean University, North Cyprus, as a full professor and was the Chairman from October 1994 to March 1996. Since March 1996, he is with the Department of Electrical and Electronics Engineering of Hacettepe University, Ankara, Turkey, where he acted as the Department Chairman during 1998-2001. He is currently the Head of the Telecommunications Group.

He conducted and supervised projects, served as a consultant and organized courses for various companies and institutions on diverse civilian and military communication systems. He served as a member of the executive committee of TUBITAK (Turkish Scientific and Technical Research Council)'s group on electrical and electronics engineering and informatics. He acted as reviewer in various national and EU projects and for distinguished journals. He was involved in the technical programme committee of many national and international conferences. He served as the Chair of 19th IEEE Conference on Signal Processing and Communications Applications (SIU 2011). He represented Turkey to COST Action 262 on Spread Spectrum Systems and Techniques in Wired and Wireless Communications. He acted as the chairman of the COST Action 289 Spectrum and Power Efficient Broadband Communications.

He was involved with high frequency asymptotic techniques, reflector antennas, wave propagation in disturbed SATCOM links, design and analysis of military SATCOM systems and spread spectrum communications. His recent research interests include multi-carrier communications, channel modelling, cooperative communications, cognitive radio and MIMO systems.

Homayoun Nikookar is an Associate Professor in the Microwave Sensing Signals and Systems Group of Faculty of Electrical Engineering, Mathematics and Computer Science at Delft University of Technology. He has received several paper awards at international conferences and symposiums and the 'Supervisor of theYear Award' at Delft University of Technology in 2010. He is the Secretary of the scientific society on Communication, Navigation, Sensing and Services (CONASENSE). He has published more than 130 refereed journal and conference papers, coauthored a textbook on 'Introduction to Ultra Wideband for Wireless Communications', Springer 2009, and has authored the book 'Wavelet Radio', Cambridge University Press, 2013.

Prof. dr. ir. Leo P. Ligthart, Ceng, FIET, FIEEE was born in Rotterdam, the Nether- lands, on September 15, 1946. He received an Engineer's degree (cum laude) and a Doctor of Technology degree from Delft University of Technology in 1969 and 1985, respectively.

He is Fellow of IET and IEEE.

He received Doctorates (honoris causa) at Moscow State Technical University of Civil Aviation in 1999, Tomsk State University of Control Systems and Radio Electronics in 2001 and the

Military Technical Academy Bucharest in 2010. He is academician of the Russian Academy of Transport.

In 1988 he was appointed as professor on Radar Positioning and Navigation and since 1992, he has held the chair of Microwave Transmission, Radar and Remote Sensing in the Department of Electrical Engineering, Mathematics and Computer Science, Delft University of Technology. In 1994, he founded the International Research Center for Telecommunications and Radar (IRCTR) and was the director of IRCTR until 2011. He received several awards from Veder, IET, IEEE, EuMA and others.

He is emeritus professor at the Delft University of Technology, is guest professor at ITB, Bandung and Universitas Indonesia in Jakarta and scientific advisor of IRCTR-Indonesia. He is founder and chairman of Conasense. He is member in the Board of Governors IEEE-AESS (2013–2015).

He is founding member of the EuMA, organized the first EuMW in 1998, the first EuRAD conference in 2004 and various conferences and symposia. He gave post-graduate courses on antennas, propagation and radio and radar applications. He was advisor in several scientific councils and consultant for companies.

Prof. Ligthart's principal areas of specialization include antennas and propagation, radar and remote sensing, but he has also been active in satellite, mobile and radio communications. He has published over 600 papers and 2 books.

2

Performance Analysis of the Communication Architecture to Broadcast Integrity Support Message

Ernestina Cianca [1], Bilal Muhammad [1], Mauro De Sanctis [1], Marina Ruggieri [1] and Ramjee Prasad [2]

[1]CTIF Italy (University of Rome "Tor Vergata")
[2]CTIF (Aalborg University, Denmark)

2.1 Introduction

Navigation capability is nowadays considered to be an assumed infrastructure and the knowledge of position (or any function of it such as speed, acceleration and heading) is nowadays fundamental to provide intelligent services for many different reasons.

First of all, the service itself could be location-based, meaning that the type of service to be provided and its configuration is dependent on the user position. Furthermore, the knowledge of the position could be an important element to optimize the design and performance of the other two key functions that need to be performed to provide intelligent services, which are sensing and communications [1]. For instance, the knowledge of the position of the sensors nodes is important to optimize the efficient dissemination of data through the sensor network (routing algorithms, or sleep/awake energy saving mechanisms). In other cases, the knowledge of the position of the mobile users could be used to optimize the communication protocols and the resource usage; in a context where different radio access networks are available, the knowledge of the position of the mobile terminals could be used to allocate the resources or even predict the allocation within a heterogeneous Radio Access Network (RAN) infrastructure. In the EU WHERE project for instance [2],

Convergence of Communications, Navigation, Sensing and Services, 31–50.

the objective was to combine wireless communications and navigation for the benefit of future mobile radio systems.

Nevertheless, the interaction between communication and navigation in this chapter is considered from a different point of view: how a communication infrastructure can be used to aid or to improve the performance of GNSS positioning systems? For instance, it is well known that a way to improve accuracy and reliability of GNSS is by broadcasting GNSS corrections generated from a network of ground stations (local, regional or global) to the user via various data links, mostly 3G networks or communication satellites (i.e. EGNOS). The Real Time Kinetic (RTK) or Precise Point Positioning (PPP) represents two examples of this approach. The choice of the communication links has an impact on the final performance of this augmented GNSS system. For instance, GEO satellites such as EGNOS, encounters limitations in urban and rural canyons, accentuated at high latitudes where the EGNOS GEO satellites are seen with low elevation angles. Studies have been done to assess the performance of EGNOS augmented GNSS for road applications [3].

Moreover, what is the impact on the communication network of the added load due to the need to transmit those corrections? This question could be important in future wide-area ITS services for urban users. Some studies have been carried out recently to reduce this load by using proper communication protocols and message format. Moreover, the performance assuming for instance less-frequent update of this broadcast information have been assessed in [4]. A communication infrastructure is needed also for broadcasting information for integrity support. This issue is gaining increasing attention and has not been deeply investigated yet. As it will be detailed in the next Sections, so far most of the works have focused on the integrity algorithm, assuming that the information contained in the so-called Integrity Support Message (ISM) is available within the required time. However, it is becoming increasingly important to better understand what is the impact of the communication architecture infrastructure that will have to disseminate these messages and what is the interaction between the format, content, frequency of update of this information, with and the requirements and capability of the chosen communication technologies. Some recent work has studied the design drivers for the Integrity Support Message architecture [5] [6]. The required size of ground monitoring networks, bounding methodology, and Time to ISM Alert are among the ISM architecture design drivers. Equally important is the dissemination network for the delivery of ISM to the final user. The choice of dissemination network strongly affects the underlying ISM architecture. In this Chapter we investigate the possibility to use TETRA standard developed by the European Telecommunication Standards Institute (ETSI) [7] to disseminate ISM messages to Public Regulated Services (PRS) aviation users. TETRA could represent a more robust (to jamming and spoofing) and low cost communication technology for the distribution of ISM to PRS users. We present preliminary results of end-to-end delay to deliver short ISM latency message to the state aircraft using TETRA network.

The Chapter is organized as follows: Section 2 presents the concept of integrity for aviation users, the requirements and existing integrity provision architectures; Section 3 explains more in details the Advanced RAIM algorithm for the provision of vertical guidance and presents possible ISM contents, formats and dissemination networks; Galileo PRS services main characteristics are presented in Section 4; TETRA standard and our proposed ISM dissemination architecture are presented in Section 5 and 6; in Section 7 the performance in terms of latency of the proposed architecture are presented; final remarks are drawn in Section 8.

2.2 Integrity for Aviation Users

Integrity is the measure of trust that can be placed in the correctness of the information provided by the navigation system. It includes the ability of the navigation system to provide timely alerts to navigation users when the system must not be used for the intended period of operation. The navigation system issues an alert within a given Time to Alert (TTA) when the error in the position solution exceeds a predefined Vertical Alert Limit (VAL) or Horizontal Alert Limit (HAL). In addition to integrity, other performance metrics of a navigation system are the accuracy, continuity and availability. The accuracy is the deviation of the estimated position solution from the true position solution. The continuity of a navigation system is its capability to perform its function without non-scheduled interruptions for the intended period of operation. The availability is defined, as the fraction of the time the navigation system is usable as compliance to accuracy, integrity, and continuity requirements for a given phase of flight. The Required Navigation Performance (RNP) for landing a civil aircraft [8] [9] is described in Table 2.1.

The maximum probability of integrity failure or integrity risk defines the probability that the navigation system does not issue alert to the navigation user within the given TTA given that the position solution exceeds the predefined VAL or HAL. On the other hand, continuity risk defines the unexpected loss

Table 2.1 Required Navigation Performance for landing a civil aircraft

Aircraft Phase of Flight	Accuracy (2σ or 95%)		Integrity Alert Limits (4-62σ)		Time To Alert	Maximum Probabilities of Failure	
	Vertical	Horizontal	Vertical	Horizontal		Integrity	Continuity
NPA Initial Approach Departure	N/A	0.22–0.74km	N/A	1.85–3.7km	10 – 15s	10^{-7}/hr	10^{-4}/hr
LNAV/VNAV	20m	220m	50m	556m	10s	1–2 x 10^{-7}/150s	4.8 x 10^{-6}/15s
LPV		16m					
AVP-I			35m	40m			
APV-II	8m		20m		<6s		
LPV-200	4m		35m				
Precision Approach CAT-I			10m				
Precision Approach CAT-II/III	<2.9m	<6.9m	5.3m	<17m	<2s	< 10^{-9}/150s	< 4 x 10^{-6}/15s

of navigation during the period of operation given that the navigation system was available before starting the operation.

The existing integrity architectures include Satellite Based Augmentation System (SBAS), Ground Based Augmentation System (GBAS), and Receiver Autonomous Integrity Monitoring (RAIM).

Satellite-Based Augmentation System (SBAS)

The SBAS integrity architecture includes a network of surveyed reference stations spread over continental areas, one or more central processing station, and a data link to provide integrity information and corrections to navigation users. The processing stations detect the anomalies in the signal-in-space and compute pseudorange corrections based on the measurements provided by the reference stations. The integrity and correction information is broadcast to the navigation users via the geostationary satellites. At present, SBAS supports navigation with vertical guidance down to LPV-200.

Ground Based Augmentation System (GBAS)

The GBAS integrity architecture concept is the same as that of SBAS. However, the GBAS uses fewer reference stations as opposed to SBAS and therefore provides coverage over a smaller. Typically, a GBAS installation serves a single airport and the surrounding terminal area by broadcasting corrections and error bounds to the aircraft via a VHF Data Broadcast (VDB) from the host airport. At present, GBAS supports all categories of precision approaches. Nevertheless, GBAS installations are very expensive.

Receiver Autonomous Integrity Monitoring (RAIM)

RAIM provides integrity by checking the consistency among the redundant (greater than four) satellite measurements. The fault detection and exclusion of faulty satellite measurements are handled at the navigation user receiver without the support of any external monitors. At present, RAIM provides lateral navigation and is used as a supplemental means of navigation for the En-route and Terminal phases of flight.

2.3 ARAIM

RAIM supporting vertical guidance using mainly airborne monitors is referred to as Advanced RAIM [10]. In the coming years, modernized GPS and GLONASS and new GNSS such as Galileo and BeiDou will be available for use worldwide [11]. In addition to this, the satellite will transmit on dual frequency civil signals. With the use of large number of dual frequency ranging signals accuracy will be improved by a huge margin. These tremendous improvements coupled with ARAIM will enable aircraft precision approach down to LPV-200 worldwide [12] [13] [14]. Figure-2.1. represents the worldwide ARAIM availability and Vertical Protection Level (VPL) as a function of user location. Enabling LPV-200 worldwide with minimal ground infrastructure requirement makes ARAIM a potential candidate of the future integrity architecture for both the integrity service providers and receiver manufacturers. The ARAIM enabled airborne receiver uses dual frequency code measurements to compute the weighted least square position solution. The fault detection and exclusion is performed in the on-board receiver followed by the computation of the vertical and horizontal protection levels. A comprehensive airborne receiver ARAIM algorithm is provided in [15].

2.3.1 ISM

ARAIM airborne receiver algorithm relies on error characterization that (over the period of time) might not be valid thus degrading the ARAIM performance. To achieve the attainable system performance, the ground station will provide information on the error characterization to the ARAIM airborne receiver in the form of Integrity Support Message (ISM) [16] [17].

2.3.1.1 ISM Content

The ISM includes two sets of parameters for each satellite, one to describe overbounding distributions for integrity purpose and another to describe nominal performance for continuity purpose. The content in particular refers to [17]:

σ^i_{URA}: User Range Accuracy (URA) of the Signal In Space Error (SISE) satisfying the integrity requirement

σ^i_{URE}: User Range Error (URE) of the SISE satisfying the continuity

$B^i_{\max}$: Maximum of the nominal bias of the SISE satisfying the integrity requirement

P^i_{sat}: Probability of satellite failure

P^i_{cons}: Probability of constellation wide fault

2.3.1.2 ISM Format

The ISM content is provided in the range domain (scalar values) or in the satellite domain (vector values) [5]. The bounding in the range domain is more conservative than that in the satellite domain. This is due to the fact that integrity bounds are inflated to protect any hypothetic worst user location in the satellite visibility area thus degrading continuity and availability of ARAIM. Sending the vector of values bounding the errors in the satellite domain (along-track, cross-track, and radial) rather than the scalar ones in the range domain improves the ARAIM performance in terms of availability and continuity [5]. Table. 2.2 compares the two formats i.e. ISM range domain and ISM satellite domain.

2.3.1.3 Time to ISM Alarm

Time to ISM Alarm (TIA) is defined as the time it takes for the ground monitoring network to identify an anomaly in the SISE and alert the aircraft about the anomaly [20]. Based on the TIA, two different integrity-monitoring approaches have been proposed [5] [18] [19]. The first one is the long latency ISM that aims at providing the information (as provided in Table 2.2) about the nominal behavior of the SISE to the final user. The ISM latency in this case is proposed on the order of days to weeks. On the other hand, short latency ISM aims at monitoring the slowly varying anomalies and sending this information to the user in terms of a short-term upper bound (denoted by B^i_{ub}) of the SISE [18]. On one hand, this relaxes the aviation user receiver fault detection requirements and on the other hand provides tighter position error bounds resulting in high system availability. The ISM latency in this case is on the order of minutes to hour. The two monitoring approaches can be used together (long + short latency ISM) as proposed in [18].

2.3.1.4 ISM Dissemination Network

ISM can be provided to the airborne receiver using a variety of dissemination networks such as GNSS in-band dissemination, Geostationary Earth Orbit (GEO) satellites, or ground-to-air communication link [18]. Each of these channels varies in coverage and supported data rate. The GNSS in-band

Table 2.2 Comparison of ISM formats

Parameter	ISM Range Domain (RD)	ISM Satellite Domain (SD)
σ^i_{URA}	Scalar value satisfying the integrity requirement	Vector of the along-track, cross-track, and radial satellite orbit and clock error components
σ^i_{URE}	Scalar value satisfying the continuity requirement	Not included in this format, a constant factor of 0.5 is used instead
$B^i_{\max}$	Scalar value of the maximum of the nominal bias satisfying the integrity requirement	Vector of mean values of the along-track, cross-track, and radial satellite orbit and clock error component
P^i_{sat}	Probability of satellite failure	Probability of satellite failure
P^i_{cons}	Probability of constellation wide fault	Probability of constellation wide fault

dissemination provides global coverage at 50bps while meeting the ICAO requirements for Safety of Life (SoL) communications. The GEO satellite such as SBAS provides regional coverage at 250bps certified by ICAO for SoL communications. Several SBAS (MSAS, GAGAN) will be foreseen in the future therefore global coverage can be achieved, however, there are no spare SBAS messages available for ARAIM ISM. SATCOM offers global coverage with varying data rate. In general, most of the SATCOM technologies in use by aviation are dedicated for the use of Air Traffic Communication (ATC) and Air Traffic Services (ATS). The possible use of SATCOM for the distribution of ISM still has to be investigated. ICAO certified [20] ground-to-air VHF link offers 31kbps with reduced coverage of approx. 43 kilometers radius offers an alternative means of ISM delivery to the aircraft.

The coverage of the ISM dissemination network, the rate at which the ISM is updated at the ground station, the frequency of ISM transmission to the airborne receiver, and the amount of information to be provided in ISM impacts the choice of ISM dissemination network. Different choices on the aforementioned design drivers lead to different ISM architecture [6] [17].

2.4 Galileo Public Regulated Service (PRS)

The key weakness of the open GPS signal today is its vulnerability to 'jamming'. A jammer is a simple device that sends out a signal at the same frequency as the satellite signal, essentially drowning out any useful information. A relatively small and inexpensive jammer, now available on the open market, can easily disrupt the open GPS signal in a limited area. More powerful jammers could disrupt signals in close proximity of airports, for example. This possibility has led to fears that terrorists could use such devices to disrupt air traffic, with severe safety and economic consequences. In addition to jamming, a second, more sinister method of GPS disruption exists, called 'spoofing'. Here, a false GPS signal is created that passes as a real GPS signal. The intended target sees what appears to be a genuine signal and is unaware that it is wrong. Both the jamming and spoofing threats pose a potential risk for aviation community and also for critical infrastructure relying on the use of GNSS.

The Galileo PRS is a dual frequency and encrypted navigation service designed to be more resistant to jamming, involuntary interference and spoofing [21]. It is similar to other Galileo services, but with some important differences: 1- Galileo PRS ensures continuity of service to authorized users when access to other navigation services is denied. 2- In cases of malicious interference, the Galileo PRS increases the likelihood of continuous availability of the Signal-in-Space (SIS).

In addition to protection against jamming and spoofing, integrity will also be provided to PRS users. The users of PRS are diverse; we refer to state aircraft (defined as aircraft used by the military, customs and police services) using Galileo PRS for vertical guidance. The RNP for PRS users [21] can be met by employing advanced Receiver Autonomous Integrity Monitoring (RAIM).

2.5 Terrestrial Trunked Radio (TETRA)

The TETRA Standard developed by the European Telecommunication Standards Institute (ETSI), defines a digital mobile radio system for voice and data communication in Ultra High Frequency (UHF) band. TETRA system offers a robust and professional mobile communication through cell type system deployments. It is a highly reliable system suitable for use by different types of users that require their communication system to provide instant transmission, high reliability and availability with an additional set of services designed to accommodate the everyday use of the system as well as provide unconditional support in extraordinary circumstances.

TETRA is a digital system, based on a combination of FDMA (Frequency division Multiple Access) and TDMA multiple access techniques. It provides four time slots on a single frequency carrier of 25 kHz thus enabling the provision of system facilities to numerous users through highly efficient use of allocated spectrum.

TETRA system incorporates mobile cell telephony types of services, traditional Private Mobile Radio (PMR) services and data transmission into a single communication system. It is therefore possible to use a single terminal to access various types of services (including data communication service), all with retaining a high level of security and privacy of communication. The most interesting aspect of data communication in the TETRA system is packet data transmission (using TETRA Packet Data Service). This allows for fully functional radio communication based on IP protocol. Packet data transmission allows the time slots to be combined in order to provide higher data throughput. Depending on the number of time slots combined, and the level of protection in the system, maximum capacity of 28.8 kbps can be achieved, as shown in Table 2.3 [22].

Table 2.3 TETRA (Release 1) multi-slot data rate

	Data throughput (kbps)		
Time Slots	High Level of Protection	Medium Level of Protection	No Protection
1	2.4	4.8	7.2
2	4.8	9.6	14.4
3	7.2	14.4	21.6
4	9.6	19.2	28.8

2.6 Distribution of ISM Using Tetra

We investigate the use of TETRA for the distribution of short latency ISM (B^i_{ub}) within the TETRA service area. We refer to the TETRA service area as "core region". We assume that the underlying ARAIM ISM architecture uses combination of long and short ISM latency. The long ISM information is provided either at state aircraft dispatch or using the Galileo I/NAV message. The state aircraft upon entering into core region will receive short latency ISM at regular intervals.

Our choice of TETRA for short latency ISM distribution to PRS users is motivated by the following reasons:

1. TETRA offers security against spoofing and malicious jamming, therefore, offering a robust mechanism for the distribution of ISM to PRS users.
2. TETRA supports Air-Ground-Air (AGA) operation, a radio service designed to provide communication at the flying speed up to 300 km/h between radio users operating from airborne assets and ground based operatives [23]. The AGA provides service by deploying an overlay network of Radio Cells or "Air Cells" that enables the communications typically from 500 feet (150 m) upwards. The use of AGA provides air-interface mechanism to distribute ISM to the state aircraft.
3. The already installed TETRA transceiver in the state aircraft for secure communication can be used to deliver short latency ISM to PRS users thus maximizing the reuse of existing infrastructure.

2.7 Results and Analysis

We assume ARAIM operation using dual constellation i.e. GPS and Galileo. The total number of satellites are 51, assuming 24 GPS and 27 Galileo satellites. The short latency ISM is transmitted every 6 minutes (360 seconds).

2.7.1 Simulation Environment

We assess the performance of TETRA in terms of end-to-end delay for short latency ISM. As shown in Figure 2.1, the gateway to dissemination network is connected with the TETRA base station using a high-speed Internet link. In this case, the communication path between the IPF and TETRA BS does not incur considerable latency, however, the dissemination latency from the TETRA BS to the state aircraft is crucial and analysed in this study. To carry out the performance assessment, we set up a simulation environment in Network Simulator 2 (NS2) [24]. The default simulation parameters are listed in Table 2.4.

2.7.2 Short Latency ISM- Range Domain

Packet size = 64 x (24 + 27) = 64 x 51 = 408 bytes

Here, 64 bits are allocated for B^i_{ub} when bounding information is provided in range domain. It can be seen in Figure-2.2 that the end-to-end short latency ISM delay increases by increasing the number of users within the cell, which is obvious. The key point here is that the short latency ISM-Range domain end-to-end delay is acceptable for the ISM update rate less than minutes 6. This implies that near real time rapid bounding, which might be needed for the precision approach (LPV-200), can be performed.

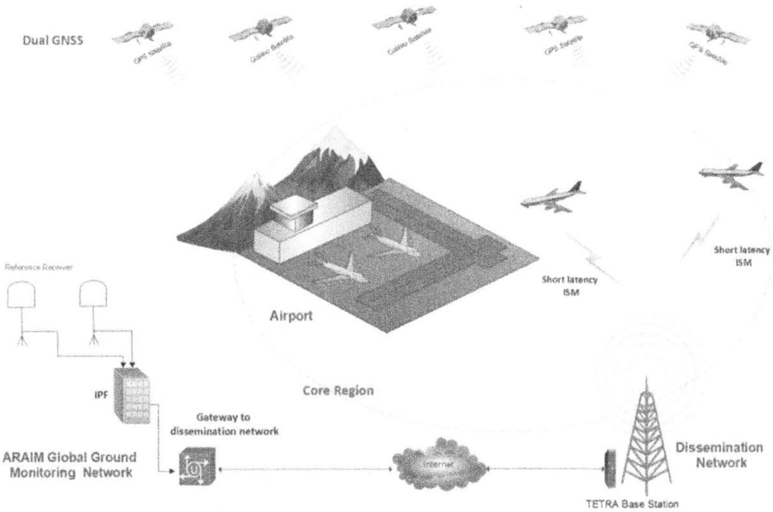

Figure 2.1 Distribution of Short Latency ISM using TETRA

2.7.3 Short Latency ISM- Satellite Domain

Packet size = 3 x 64 x (24 + 27) = 1224 bytes

In this case, 3 x 64 bits are allocated B^{i}_{ub} when bounding information is provided in satellite domain. The packet size increased approx. 3 times in case of sending the bounding information in satellite domain. As can be seen in Figure-2.3, even when the packet size is increased, compatible values can be reached for short latency ISM update rate on the order of 2–6 minutes.

2.8 Conclusions

In the coming years, the use of existing telecommunication networks including both terrestrial and satellite networks will play a major role in the effectiveness of ARAIM architecture. Important design drivers of the ARAIM architecture are related to the reuse and/or reduction of the infrastructure (including ground-monitoring networks and dissemination network) needed for providing worldwide vertical guidance using ARAIM. In this framework, this chapter presents preliminary results related to the use of TETRA standard to disseminate short latency ISM messages to Public Regulated Services (PRS) aviation users. These results show that both in case of range-domain or satellite-domain, the end-to-end delay for short latency ISM can be met. However, further studies are needed using different bit error rates according to ICAO BER requirements, sending ground validated set of precise navigation information within ISM, and end-to-end short latency ISM delay using high data rate (i.e. TEDS).

Table 2.4 Simulation Parameters

Parameter Description	Value
Number of cell (s)	1
Number of users per cell	5, 10, 15, 20, 25, 30
Bit Error Rate (BER)	10^{-6}
Transport Agent	TCP
Data rate	19.2 kbps
Simulation time	12 hours
Channel access scheme	TDMA
Maximum number of slots per user	4
Slot time	14.167 ms
Slot length (bytes)	64 bytes

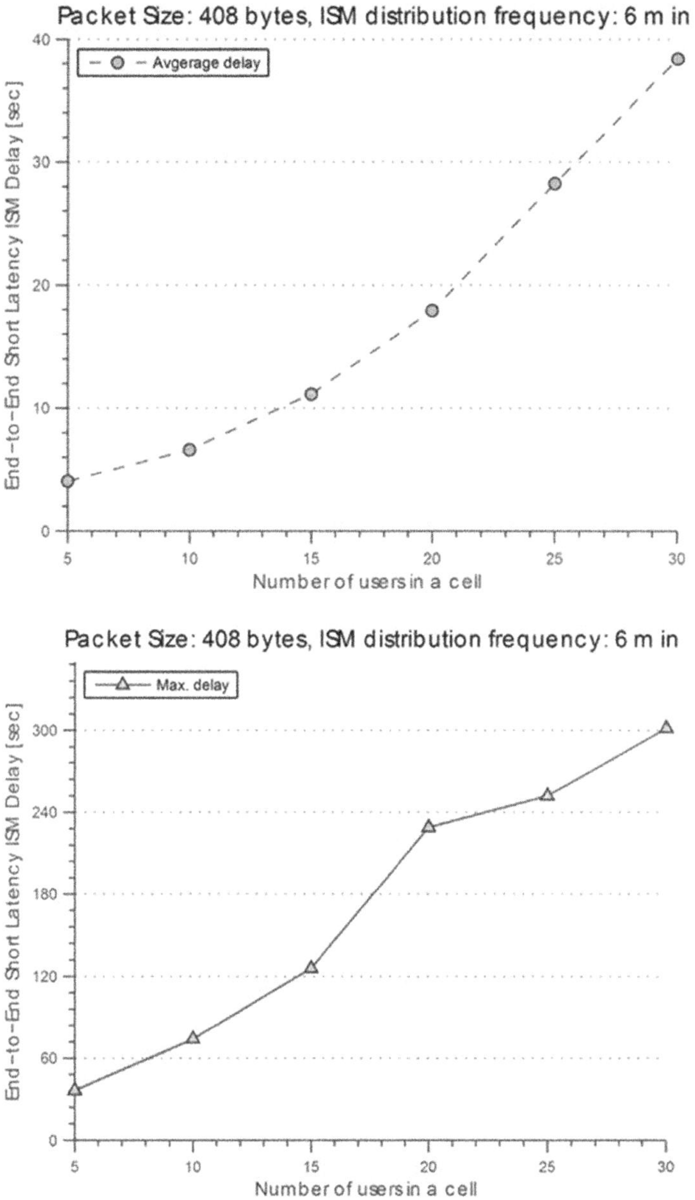

Figure 2.2 Average and Maximum end-to-end Short Latency ISM (RD) Delay for Different Number of Users in a Cell

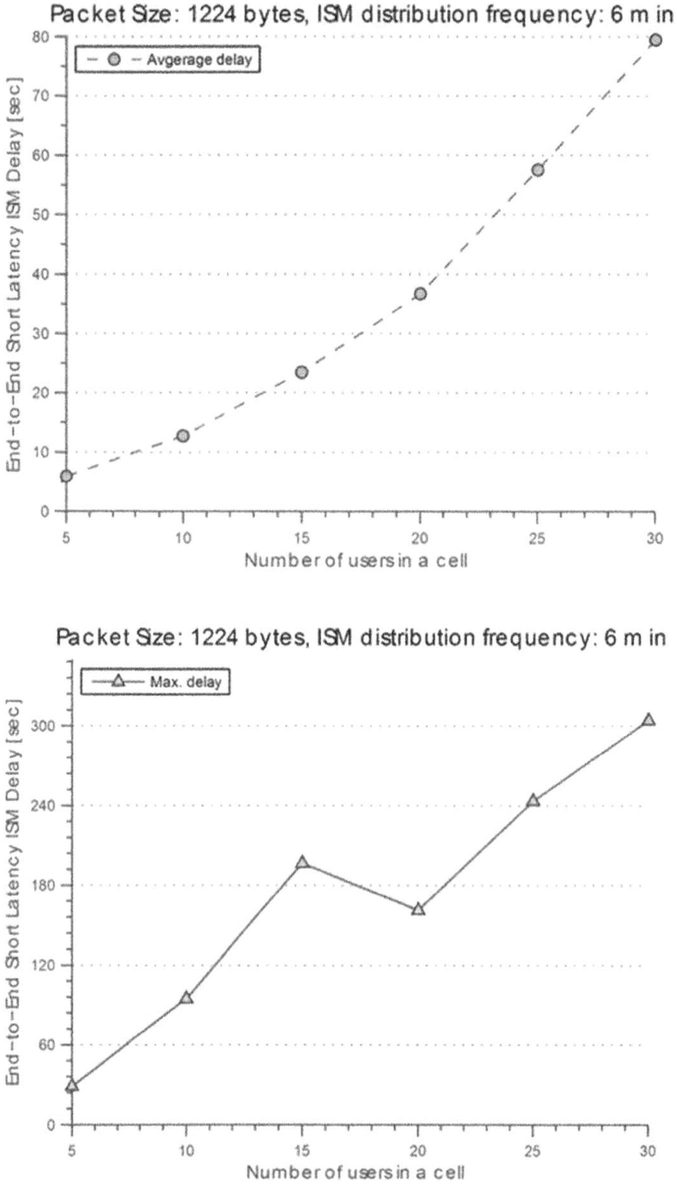

Figure 2.3 Average and Maximum end-to-end Short Latency ISM (SD) Delay for Different Number of users in a Cell

References

[1] "Communications, Navigation, Sensing and Services (CONASENSE)", River Publishers, Editors: L.P. Ligthart, & R. Prasad.

[2] Http://www.ict-where.eu/

[3] M. Obst, R. Schubert, R. Streiter, C. Libeerto, "Benefit Analysis of EGNOS/EDAS for Urban Road Transport Applications," Proc. of the European Conference on ITS, Lyon, 2011.

[4] Ming Qu, "Experimental Studies of Wireless Communication and GNSS Kinematic Positioning Performance in High-Mobility Vehicle Environments," Master Thesis, Faculty of Science and Engineering, Queensland University of Technology, March 2012.

[5] I. Martini, M. Rippl, and M. Meurer, Integrity Support Message Architecture Design for Advanced Receiver Autonomous Integrity Monitoring," Proceedings of European Navigation Conference, Vienna Austria, 23–25 April 2013.

[6] T. Walter, J. Blanch, and P. Enge, "A Framework for Analyzing Architectures that Support ARAIM," Proceedings of the 25th International Technical Meeting of The Satellite Division of the Institute of Navigation (ION GNSS 2012, (Nashville, TN), September 2012.

[7] ETSI TR 102 300–6, "Terrestrial Trunked Radio (TETRA); Voice plus Data (V+D); Designers' guide; Part 6: Air-Ground-Air," tech. rep., 2011.

[8] ICAO, Annex 10 to the Convention of International Civil Aviation, Montreal, PQ, Canada, July 17, 2007, vol. 1, Radio Navigation Aids, Amendment 82.

[9] Ene, Alexandru, "Utilization of modernized global navigation satellite systems for aircraft-based navigation integrity", ProQuest, 2009.

[10] J. Blanch, T. Walter, P. Enge, S. Wallner, F. A. Fernandez, R. Dellago, R. Ionnides, B. Pervan, I. F. Hernandez, B. Belabbas, A. Spletter, and M. Rippl, "A Proposal for Multi-constellation Advanced RAIM for Vertical Guidance," Proceedings of the 24th International Technical Meeting of The Satellite Division of the Institute of Navigation (ION GNSS 2011), Portland, OR, September 2011, pp. 2665–2680.

[11] Hegarty, Christopher J., and Eric Chatre. "Evolution of the global navigation satellitesystem (gnss)." Proceedings of the IEEE 96.12 (2008): 1902–1917.

[12] T. Walter, J. Blanch, P. Enge, B. Pervan, and L. Grattan, "Future Architectures to Provide Aviation Integrity," Proceedings of the 2008 National

Technical Meeting of the Institute of Institute of Navigation, San Diego, CA, January 2008, pp. 394–401.

[13] Todd Walter, P. Juan Blanch, and Boris Pervan, "Worldwide Vertical Guidance of Aircraft Based on Modernized GPS and New Integrity Augmentations," Proceedings of IEEE, vol. 96, no. 12, pp. 1918–1935, 2008.

[14] Y. C. Lee and M. P. McLaughlin, "Feasibility Analysis of RAIM to Provide LPV-200 Approaches with Future GPS," Proceedings of the 20th International Technical Meeting of the Satellite Division of The Institute of Navigation (ION GNSS 2007), Fort Worth, TX, September 2007.

[15] J. Blanch, T. Walter, P. Enge, Y. Lee, B. Pervan, M. Rippl, and A. Spletter, "Advanced RAIM User Algorithm Description: Integrity Support Message Processing, Fault Detection, Exclusion, and Protection Level Calculation," Proceedings of the 2012 Global Navigation Satellite Systems Conference of the Institute of Navigation (ION GNSS 2012), Nashville, USA, September 2012.

[16] Federal Aviation Administration (FAA), " Phase-II of the GNSS Evolutionary Architecture Study," Feburary 2010.

[17] EU-US ARAIM Technical Sub Group of the Working Group C, "GPS-Galileo Working Group C ARAIM Technical Subgroup Interim Report (Issue 1.0, 19 December 2012)," tech. rep., 2012.

[18] I. Martini, M. Rippl, and M. Meurer, "Advanced RAIM Architecture Design and User Algorithm Performance in a real GPS, GLONASS, and Galileo Scenario" Proceedings of the Proceedings of the 26th International Technical Meeting of The Satellite Division of the Institute of Navigation (ION GNSS+ 2013) September 16 – 20, 2013, Nashville, TN.

[19] J. Blanch, T. Walter, and P. Enge, "Advanced RAIM System Architecture with a Long Latency Integrity Support Message," Proceedings of the Proceedings of the 26th International Technical Meeting of The Satellite Division of the Institute of Navigation (ION GNSS+ 2013), September 16 – 20, 2013, Nashville, TN.

[20] ICAO, Annex 10 to the Convention of International Civil Aviation, Second edition, July 2007, vol. 3, Communication Systems.

[21] European Commission, "Galileo Mission High Level De_nition," 23 September 2002.

[22] Bakaric, Shiga, et al. "TETRA (terrestrial trunked radio)-technical features and application of professional communication technologies in mobile digital radio networks for special purpose services." Proceedings

of 47th International Symposium ELMAR-2005, 08–10 June 2005, Zadar, Croatia.

[23] ETSI TS 100 392-18–4, "Terrestrial Trunked Radio (TETRA); Voice plus Data (V+D) and Direct Mode Operation (DMO); Part 18: Air interface optimized applications; Sub-Part 4: Net Assist Protocol 2," tech. rep., 2012.

[24] Kevin Fall, Kannan Varadhan, ed., The ns Manual. August 24 2000.

Biographies

Ernestina Cianca received the Laurea degree in Electronic Engineering "cum laude" at the University of L'Aquila in 1997. She got the Ph.D. degree at the University of Rome Tor Vergata in 2001. She concluded her Ph.D. at Aalborg University where she has been employed in the Wireless Networking Groups (WING), as Research engineer (2000–2001) and as Assistant Professor (2001–2003). Since Nov. 2003 she is Assistant Professor in Telecommunications at the URTV (Dpt. of Electronics Engineering), teaching DSP, Information and Coding Theory and Advanced Transmission Techniques. She is the co-director of a II level Master in Advanced Satellite Communication and Navigation Systems. She has been the principal investigator of the WAVE-A2 mission, funded by the Italian Space Agency and aiming to design payloads in W-band for scientific experimental studies of the W-Band channel propagation phenomena and channel quality. She has been coordinator of the scientific activities of the Electronic Engineering Department on the following projects: ESA project European Data Relay System (EDRS); feasibility study for the scientific small mission FLORAD (Micro-satellite FLOwer Constellation of millimeter-wave RADiometers for the Earth and space Observation at regional scale); TRANSPONDER2, funded by ASI, about the design of a payload in Q-band for communications; educational project funded by ASI EduSAT on pico-satellites. She has worked on several European and National projects. Her research mainly concerns wireless access technologies (CDMA and MIMO-OFDM-based systems), integration of terrestrial and satellite systems, short-range communications in biomedical applications. She has been General Chair of the conference ISABEL 2010 (Third Symposium on Applied Sciences in Biomedical and Telecommunication Engineering), she has been TPC Co-Chair of the conference European Wireless Technology 2009 (EuWIT2009); TPC Co-Chair in the conference Wireless Vitae 2009. She is Guests Editors of some Special Issues in journals such as Wireless Personal Communications (Wiley) and Journal of Communications (JCM, ISSN 1796–2021). She is author of about 70 papers, on international journals/transactions and proceedings of international conferences.

Bilal Muhammad obtained his bachelor's degree in Computer Engineering in 2005 from the COMSATS Institute of Information Technology, Abbottabad, Pakistan. He carried out his master thesis at the Ericsson Research, Lulea, Sweden, and received his master's degree in Electrical Engineering specialized in Telecommunication from Blekinge Institute of Technology (BTH), Sweden, in 2008. Currently, he is a Ph.D. candidate at the Ph.D. school of Telecommunication and Microelectronics, University of Rome Tor Vergata. He is working on the design and performance analysis of Advanced Receiver Autonomous Integrity Monitoring (ARAIM) architecture. His research work is sponsored by the Italian Space Agency and the Italian Air Traffic Service company ENAV under the SENECA (**S**at**E**llite **N**avigation s**E**rvices for **C**ivil **A**viation) program.

Mauro De Sanctis received the "Laurea" degree in Telecommunications Engineering in 2002 and the Ph.D. degree in Telecommunications and Microelectronics Engineering in 2006 from the University of Roma "Tor Vergata" (Italy). He was with the Italian Space Agency (ASI) as holder of a two-years research fellowship on the study of Q/V band satellite communication links for a technology demonstration payload, concluded in 2008; during this period he participated to the opening and to the first trials of the ASI Concurrent Engineering Facility (ASI-CEF). From the end of 2008 he is Assistant Professor at the Department of Electronics Engineering, University of Roma "Tor Vergata" (Italy), teaching "Information and Coding Theory". From January 2004 to December 2005 he has been involved in the MAGNET (My personal Adaptive Global NET) European FP6 integrated project and in the

SatNEx European network of excellence. From January 2006 to June 2008 he has been involved in the MAGNET Beyond European FP6 integrated project as scientific responsible of WP3/Task3. He has been involved in research activities for several projects funded by the Italian Space Agency (ASI): DAVID satellite mission (DAta and Video Interactive Distribution) during the year 2003; WAVE satellite mission (W-band Analysis and VErification) during the year 2004; FLORAD (Micro-satellite FLOwer Constellation of millimeter-wave RADiometers for the Earth and space Observation at regional scale) during the year 2008; CRUSOE (CRUising in Space with Out-of- body Experiences) during the years 2011/2012. He has been involved in several Italian Research Programs of Relevant National Interest (PRIN): SALICE (Satellite-Assisted Locallzation and Communication systems for Emergency services), from October 2008 to September 2010; ICONA (Integration of Communication and Navigation services) from January 2006 to December 2007, SHINES (Satellite and HAP Integrated NEtworks and Services) from January 2003 to December 2004, CABIS (CDMA for Broadband mobile terrestrial-satellite Integrated Systems) from January 2001 to December 2002. In 2007 he has been involved in the Internationalization Program funded by the Italian Ministry of University and Research (MIUR), concerning the academic research collaboration of the Texas A&M University (USA) and the University of Rome "Tor Vergata" (Italy). He is currently involved in the coordination of scientific activities of the experiments for broadband satellite communications in Q/V band (Alphasat Technology Demonstration Payload 5 - TDP5) funded jointly by ASI and ESA. He is serving as Associate Editor for the Space Systems area of the IEEE Aerospace and Electronic Systems Magazine. His main areas of interest are: wireless terrestrial and satellite communication networks, satellite constellations (in particular Flower Constellations), resource management of short range wireless systems. He co-authored more than 60 papers published on journals and conference proceedings. He was co-recipient of the best paper award from the 2009 International Conference on Advances in Satellite and Space Communications (SPACOMM 2009).

Ramjee Prasad is currently the Director of the Center for TeleInfrastruktur (CTIF) at Aalborg University, Denmark and Professor, Wireless Information Multimedia Communication Chair.

Ramjee Prasad is the Founding Chairman of the Global ICT Standardisation Forum for India (GISFI: www.gisfi.org) established in 2009. GISFI has the purpose of increasing of the collaboration between European, Indian, Japanese, North-American and other worldwide standardization activities in the area of Information and Communication Technology (ICT) and related application areas. He was the Founding Chairman of the HERMES Partnership – a network of leading independent European research centres established in 1997, of which he is now the Honorary Chair. He is the founding editor-in-chief of the Springer International Journal on Wireless Personal Communications. He is a member of the editorial board of other renowned international journals including those of River Publishers. Ramjee Prasad is a member of the Steering, Advisory, and Technical Program committees of many renowned annual international conferences including Wireless Personal Multimedia Communications Symposium (WPMC) and Wireless VITAE. He is a Fellow of the Institute of Electrical and Electronic Engineers (IEEE), USA, the Institution of Electronics and Telecommunications Engineers (IETE), India, the Institution of Engineering and Technology (IET), UK, and a member of the Netherlands Electronics and Radio Society (NERG), and the Danish Engineering Society (IDA). He is a Knight ("Ridder") of the Order of Dannebrog (2010), a distinguished award by the Queen of Denmark.

3

Nodes Selection for Distributed Beamforming (DB) in Cognitive Radio (CR) Networks

X. Lian[1], H. Nikookar[1] and L. P. Ligthart[1]

[1]Department of Electrical Engineering, Mathematics and Computer Science, Microwave Sensing, Signals & Systems, Delft University of Technology, The Netherlands

In this chapter, we introduce the DB technique to CR networks, which are constituted of distributed CR nodes. The goal of the DB method is to collaboratively forward the CR signal to the Distant CR (DCR) user. To solve the problem of the extreme narrow main beam in the pattern when we introduce the DB method into CR networks, we propose a novel Nodes Selection (NS) method. The presented NS method is based on the differences in beam width of a broadside array and an end-fire array. We select those CR nodes, which are able to form a full 0size end-fire array and a reduced size broadside array. This NS method chooses those CR nodes which are located in the "belt" area along the direction of the DCR user. Simulation results of the average beampattern of our NS method show that the main beams are successfully directed towards the DCR users and are enlarged for practical applications in CR networks. What is more, for a CR network with large physical size, our NS method can widen the main beam while maintaining sufficiently low sidelobe levels for CR transmission.3

3.1 Introduction

Many appealing applications of wireless communications have been emerging, such as mobile internet access, health care and medical monitoring, smart

Convergence of Communications, Navigation, Sensing and Services, 51–70.

homes etc. With a remarkable growth in designing and manufacturing of various sensors for health care, transportation, environment monitoring, there has been an increasing demand of versatile wireless services. Another emerging trend of the current wireless services is the demand of high date rates wideband services. Thus there is a need to develop an energy efficient green technique that is capable of optimizing the premium radio resources, such as power and spectrum, while guaranteeing desirable Quality of Services (QoS). The new techniques should be designed to spatially, temporally and spectrally minimize the energy spent to transmit information to achieve high energy efficiency.

Cognitive Radio (CR) is a promising solution to meet this requirement. CR has been initially introduced by Joseph Mitola [1], who described how CR could enhance the flexibility of wireless services through a radio knowledge representation language. Though there are different existing definitions of CR, all of them deliver six keywords; they are: awareness, intelligence, learning, adaptation, reliability and efficiency. CR can be considered as a radio that is able to behave as a cognitive system, having at least the capabilities of observing, making decisions and adapting.

CR is able to utilize the unused spectrum efficiently in a dynamically changing environment. It provides various solutions to accommodate this spectrum to be used by unlicensed wireless devices without disrupting the communications of the Primary Users (PUs) of the spectrum [2]. It can also achieve efficient radio resource management while providing high date rate and reliable wireless communication services via implementation of cognitions in three domains. They are time, frequency and space domains.

In this chapter, we focus on discussing the spatial potentials of CR by limiting ourselves to the spatial diversity of the CR system. In space domain, if CR is able to transmit signals to its users while ensuring those signals to be received by PU is below the interference level of PU, CR is able to have the full access of the spectrum utilized by the PU. This helps CR to achieve the most efficient spectrum usage by totally sharing the whole spectrum with PU.

CR capabilities may also be exploited in Wireless Sensor Networks (WSN), which are traditionally assumed to employ a fixed spectrum allocation and characterized by the communication and processing resource constrains of low-end sensor nodes[3]. Depending on the applications, WSN composed of sensor nodes equipped with CR may benefit from its potential advantages, such as dynamic spectrum access and adaptability for reducing power consumptions. A CR network is formed by nodes that are geographically distributed in a certain area, which is shown in Figure 3.1. Those nodes are possibly wireless terminals, subscriber users, or sensors in the CR network. In this chapter, we

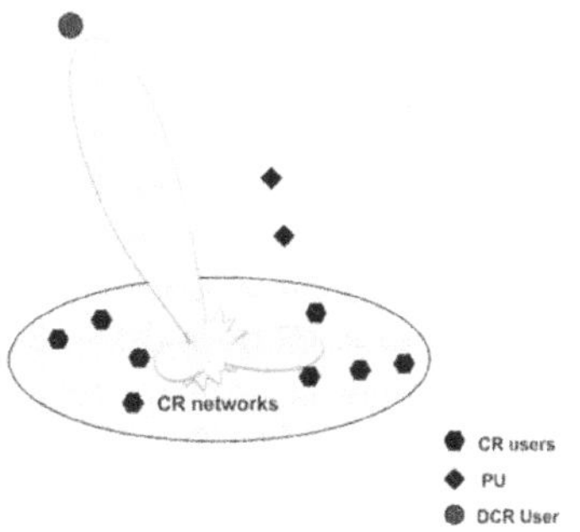

Figure 3.1 Coexistence of PU and CR with DB Techniques

propose to employ Distributed Beamforming (DB) at CR networks in order to form beams towards Distant CR (DCR) users so that the CR network is able to forward the signals to DCR users cooperatively. Via adopting the DB method, the CR networks increases its coverage range without causing harmful interferences to PU.

DB is also referred to as collaborative beamforming, and is originally employed as an energy-efficient scheme to solve long distance transmission in WSN, in order to reduce the amount of required energy and consequently to extend the utilization time of the sensors [4]. The basic idea of DB is that a set of nodes in a wireless network acts as a virtual antenna array and then forms a beam towards a certain direction to collaboratively transmit a signal. DB has been proposed in [4] and it has been shown that by employing K collaborative nodes, the collaborative beamforming can result in up to K-fold gain in the received power at a distant access point. Recently a cross-layer approach for DB in wireless ad-hoc networks has been discussed in [5] applying two communication steps. In the first phase, nodes transmit locally in a random access time-slotted fashion. In the second phase, a set of collaborating nodes, acting as a distributed antenna system, forward the received signal to one or more faraway destinations. The improved beam pattern and connectivity properties have been shown in [6], and a reasonable beamforming performance affected by nodes synchronization errors has been discussed in [7]. DB has also been introduced in relay communication systems [8–12]. Different types of relays are considered, e.g. amplify-and-forward (AF) relays, Filter and Forward (FF) relays, etc. The models of relay networks [8–12] have a source, a relay, and a destination, where transmit DB is employed both at the source and at the relay. The authors in [8–11] have developed several DB techniques for relay networks with flat fading channels. In [12], frequency selective fading has been considered.

DB requires accurate synchronization; in other words, the nodes must start transmitting at the same time, synchronize their carrier frequencies, and control their carrier phases so that their signals can be combined constructively at the destination. A synchronization technique based on time-slotted round-trip carrier synchronization has been proposed for DB in [13], and a review has been given in [14]. In this chapter, we adopt the master-slave architecture proposed in [7], where a designated master transmitter (one of the CR nodes in the networks) coordinates the synchronization of others (slave) transmitters for DB. Since this method has also been proved in [7] that a large fraction of DB gains can still be realized even with imperfect synchronization corresponding to phase errors with moderately large variance, we focus more on introducing DB methods rather than discussing the synchronization algorithms.

To apply the DB technique in CR networks, a practical difficulty arises due to the fact that the width of the main beam in the beampattern generated by the DB method is relying on the working frequency of the CR networks. As shown in [4], the main beam of the beampattern will become narrower when $\widetilde{R}$ increases, where $\widetilde{R} \triangleq R/\lambda$, R is the covering radius of the network and λ is the wavelength. If we consider a CR network with R=100m and utilizing the spectrum of the UHF band, which is, for instance, 750MHz, we can obtain that $\widetilde{R} = 250$. In [4], the authors have shown that the width of the main beam can be approximated by $\frac{35^\circ}{\widetilde{R}}$. Thus in our example, the width of the main beam will become about 0.1°. This is considered to be too narrow, implying that once the DOA estimation of the distant nodes is not accurate enough, the main beam in the pattern may miss its direction. Meanwhile it also reveals that the width of the main beam in the beampattern mostly relies on the center frequency at which the CR network is able to access.

To solve this problem of the extreme narrow main beam, in this chapter we will propose a Nodes Selection (NS) method. The presented NS method is based on the differences in beam width of a broadside array and an end-fire array. The beampatterns of these two types of arrays allow us to find out how the main beam of the end-fire array is much wider than that of the broadside array. We thus conclude that the "broadside" size of the CR network should be small so that a beampattern with a wider main beam can be maintained. As a result, we suggest selecting CR nodes, which are able to form a full size end-fire array and a reduced size broadside array. In other words, we choose those CR nodes which are located in the "belt" area along the direction of the distant nodes.

The chapter is arranged as follows. Section 1.2 introduces DB technique to CR networks. A novel NS method to enlarge the main beam of the beampattern generated by the DB technique is presented in section 1.3. Simulations will be given afterwards in section 1.4, showing that our NS method is effective in generating a wider main beam in the beam pattern. Section 1.5 concludes the whole chapter.

3.2 DB for CR Networks

The geometrical structure of the first model together with distant receiver terminals including PU and DCR users is illustrated in Figure 3.2. K CR nodes are uniformly distributed over a disc centered at O with radius R. We denote the polar coordinates of the kth CR node by (r_k, Ψ_k). The number of DCR users is L_{DCR}. These DCR users are considered as access points, and located in the same plane at$(A_i^{DCR}, \phi_i^{DCR})$, $i = 1, 2, ..., L_{DCR}$. Meanwhile the number of PUs is L_{PU}, and these users coexist with DCR users. Their locations are (A_i, ϕ_i), $i = 1, 2, ..., L_{PU}$. The CR nodes in the CR network are requested to form a virtual antenna array and collaboratively transmit a common message$s(t)$.

3.2.1 Necessary Assumptions

Without loss of generality, we adopt the following assumptions:

1) The number of CR nodes are larger than that of DCR users plus the number of PUs, i.e., $K > L_{DCR} + L_{PU}$. This is basically required

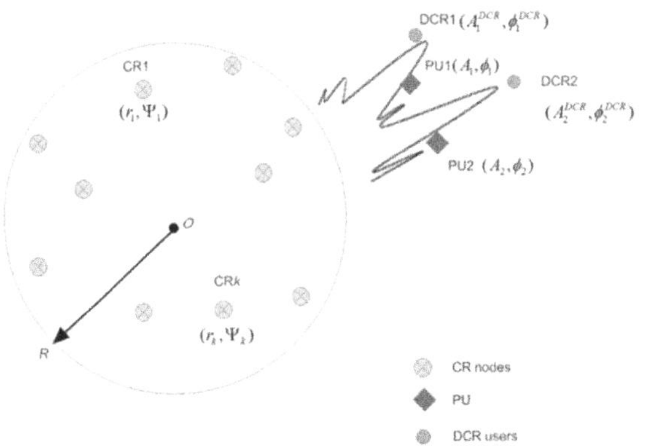

Figure 3.2 CR networks with DCR and PU

to solve the latter matrix equations in this chapter. Many Adaptive Beamforming techniques also request that the number of array antennas is larger than that of the constraint directions for main beam and null patterns.

2) All DCR users and PUs are located in the far field of the CR network, such that $A_i^{DCR} >> R,\ i = 1, 2, ..., L_{DCR}$ and $A_i >> R,\ i = 1, 2, ..., L_{PU}$.
3) The bandwidth of $s(t)$ is narrow, so that $s(t)$ is considered to be constant during the time interval R/c, where c is the speed of light.

3.2.2 DB for CR Networks

Let $x_k(t)$ denotes the transmitted signal from the kth node,

$$x_k(t) = s(t)e^{j2\pi ft} \tag{3.1}$$

where f is the carrier frequency. The received signal at an arbitrary point (A, ϕ)in the far field due to the kth node transmission is [15]

$$r_k(t) = \beta_k x_k(t - \frac{d_k}{c}) = \beta_k s(t - \frac{d_k}{c})e^{j2\pi ft}e^{-j\frac{2\pi}{\lambda}d_k} \tag{3.2}$$

where d_k is the distance between the kth node and the access point (A, ϕ), and $\beta_k = (d_k)^{\frac{-\gamma}{2}}$ is the signal path loss with γ donating the path loss exponent. Making use of assumption 2 in the previous paragraph [15], β_k and $d_k$are approximated by

$$d_k = \sqrt{A^2 + r_k^2 - 2Ar_k\cos(\phi - \Psi_k)} \approx A - r_k\cos(\phi - \Psi_k) \tag{3.3}$$

$$\beta_k = (d_k)^{\frac{-\gamma}{2}} \approx [A - r_k\cos(\phi - \Psi_k)]^{\frac{-\gamma}{2}} \approx \beta\left(1 + \frac{\gamma r_k\cos(\phi - \Psi_k)}{2A}\right) \tag{3.4}$$

where $\beta = A^{\frac{-\gamma}{2}}$. It is also ensured that $\frac{\gamma r_k\cos(\phi - \Psi_k)}{2A} << 1$. Thus β_k can then be approximated by β. Substituting equation (3.3) and (3.4) into equation (3.2), it follows that [15],

$$r_k(t) \approx \beta e^{-j\frac{2\pi}{\lambda}A}s(t - \frac{A}{c})e^{j2\pi ft}e^{j\frac{2\pi}{\lambda}r_k\cos(\phi - \Psi_k)} \tag{3.5}$$

We assume there is only one DCR, i.e., $L_{DCR} = 1$, and simplify $(A_1^{DCR}, \phi_1^{DCR})$by (A_0, ϕ_0). The case with more than one DCR users has been discussed in [16]. As proposed in [4], we adopt for the initial phase of each node

$$\varphi_k = -\frac{2\pi}{\lambda} r_k \cos(\phi_0 - \Psi_k) \tag{3.6}$$

The received signal $r_k(t)$ at (A, ϕ) becomes

$$r_k(t) \approx \beta e^{-j\frac{2\pi}{\lambda}A} s(t - \frac{A}{c}) e^{j2\pi f t} e^{j\frac{2\pi}{\lambda} r_k \cos(\phi - \Psi_k)} e^{-j\frac{2\pi}{\lambda} r_k \cos(\phi_0 - \Psi_k)} \tag{3.7}$$

The array factor $F(\phi|\, r_k, \Psi_k)$ yields:

$$\begin{aligned} F(\phi|\, r_k, \Psi_k) &\approx \frac{1}{K}\sum_{k=1}^{K} e^{j\frac{2\pi}{\lambda} r_k [\cos(\phi - \Psi_k) - \cos(\phi_0 - \Psi_k)]} \\ &= \frac{1}{K}\sum_{k=1}^{K} e^{-j\frac{4\pi}{\lambda} r_k \sin(\frac{\phi - \phi_0}{2}) \sin(\frac{\phi + \phi_0 - 2\Psi_k}{2})} \end{aligned} \tag{3.8}$$

We assume there are many CR nodes and the locations of CR nodes follow a uniform distribution over the disk of radius R, leading to the probability density functions (pdf)

$$\begin{cases} f_{r_k}(r) = \frac{2r}{R^2}, \ 0 \le r < R \\ f_{\Psi_k}(\Psi) = \frac{1}{2\pi}, \ -\pi \le \Psi_k < \pi \end{cases} \tag{3.9}$$

If the CR nodes are distributed according a two-dimensional Gaussian process, the corresponding DB techniques and its beampattern characteristics can be found in [17].

By defining $z_k \triangleq \frac{r_k}{R} \sin\left(\Psi_k - \frac{\phi_1 + \phi_0}{2}\right)$, the compound random variable z_k has a pdf [4]

$$f_{z_k}(z_k) = \frac{2}{\pi}\sqrt{1 - z_k^2}, \ -1 \le z < 1 \tag{3.10}$$

The array factor in equation (3.8) can now be written as

$$F(\phi\,|z_k) = \frac{1}{K}\sum_{k=1}^{K} \exp\left(-j4\pi\widetilde{R}\sin\left(\frac{\phi - \phi_0}{2}\right) z_k\right) \tag{3.11}$$

where $\widetilde{R} \triangleq R/\lambda$ is the radius of the disk normalized by the wavelength. The far field beampattern is defined by

$$P(\phi\,|z_k) \triangleq |F(\phi\,|z_k)|^2 \tag{3.12}$$

and the average array beam pattern of the CR networks becomes [4]

$$P_{av}(\phi) \triangleq \mathrm{E}\left[P(\phi\,|z_k)\right] = \frac{1}{K} + \left(1 - \frac{1}{K}\right)\mu^2(\phi) \tag{3.13}$$

where

$$\mu(\phi) = E\left[F(\phi)\right] = \left|\frac{2J_1\left(\alpha(\phi)\right)}{\alpha(\phi)}\right| \tag{3.14}$$

$$\alpha(\phi) \triangleq 4\pi\widetilde{R}\sin\left(\frac{\phi - \phi_0}{2}\right) \tag{3.15}$$

$J_n(\cdot)$ stands for the *n*th order Bessel function of the first kind, which is defined by $J_n(x) = \sum\limits_{k=0}^{\infty} \frac{(-1)^k x^{n+2k}}{k!(n+k)!2^{n+2k}}$ and $E[\cdot]$ stands for the statistical expectation. Above equations learn us that if each CR node adopts the initial phase as given in equation (3.6), the average pattern generated by the whole CR network can be obtained from equation (3.13).

Figure 3.3 shows the average beampattern of the DB method for different K ($K = 8, 16$) and $\widetilde{R}(\widetilde{R} = 2, 4, 8)$. We assume that the direction of DCR

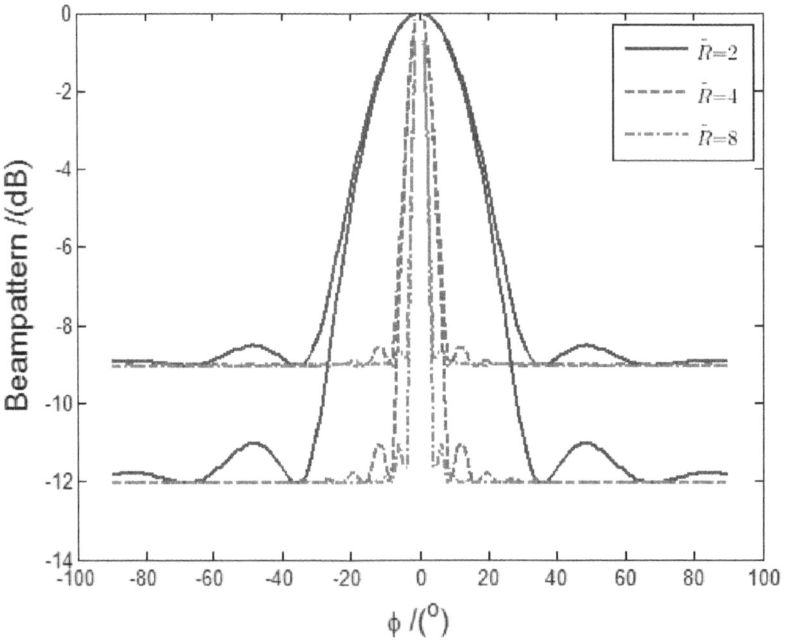

Figure 3.3 Average Beampattern of the DB Method

is $\phi_0 = 0°$. It can be seen that when the beam angle moves away from the direction of DCR, the sidelobe approaches $\frac{1}{K}$, i.e., $10\log_{10}\left(\frac{1}{8}\right) \approx -9dB$and $10\log_{10}\left(\frac{1}{16}\right) \approx -12dB$, respectively. This leads to the logical conclusions that the sidelobe level decreases when K increases, and that the larger $\widetilde{R}$ becomes, the narrower the main beam will be, and consequently the better directivity in the beampattern will be achieved. Figure 3.3 is only for demonstration purposes, because the value of the parameters that we considered ($K = 8, 16$; $\widetilde{R} = 2, 4, 8$) are not realistic in real application. In practical applications, $\widetilde{R}$ should be large enough so that CR networks can have enough CR nodes.

3.3 NS for CR Networks with Enlarged Main Beam

We have mentioned in the introduction that $\widetilde{R}$ increases rapidly when the working frequency of the CR networks goes higher. In this section we propose a new NS method for CR networks to select proper nodes to achieve wider main beam in the beampattern of the DB method.

We first consider two extreme cases of the CR network, which are two types of array antennas: broadside array and end-fire array by projecting the location of each CR node along an X and Y axis, and study the properties of these two array antennas. After this we propose a NS method.

We assume one DCR user is located along the X axis ($\phi_0 = 0°$). We then consider the location of a CR node by projecting it into the Cartesian coordinate system (X and Y directions) as shown in Figure 3.4. In this way we create two virtual arrays: broadside array and end-fire array. We now discuss the performance of the two separate arrays (broadside array and end-fire array) instead of the full CR network.

The average beampatterns of these two arrays are summarized in the following equations

$$\overline{P}_{broadside}(\phi) = \frac{1}{K} + \left(1 - \frac{1}{K}\right)\mu_b^2(\phi) \tag{3.16}$$

where

$$\mu_b(\phi) = \left|\frac{2J_1(\alpha_b(\phi))}{\alpha_b(\phi)}\right| \tag{3.17}$$

$$\alpha_b(\phi) = 2\pi\widetilde{R}(\sin\phi - \sin\phi_0) \tag{3.18}$$

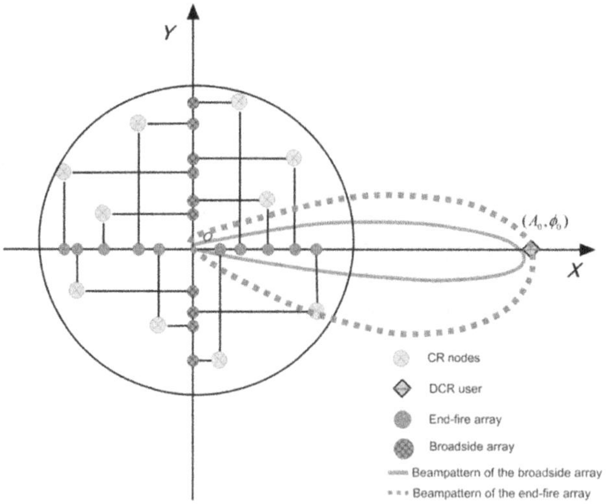

Figure 3.4 Converting Locations of CR Nodes into Broadside and End-Fire Arrays

and

$$\overline{P}_{end-fire}(\phi) = \frac{1}{K} + \left(1 - \frac{1}{K}\right) \mu_e^2(\phi) \tag{3.19}$$

where

$$\mu_e(\phi) = \left| \frac{2J_1(\alpha_e(\phi))}{\alpha_e(\phi)} \right| \tag{3.20}$$

$\alpha_e(\phi) = 2\pi\widetilde{R}(\cos\phi - \cos\phi_0)$ Proofs of equations (3.16) to (3.21) can be found in Appendix A.

The results of equation (3.16) and (3.21) are shown in Figure 3.5. We assume there are 32 nodes in the CR network and the normalized radius is 35, i.e., $K = 32$ and $\widetilde{R} = 35$. We also assume there is only one DCR user, and its DOA is $\phi_0 = 0°$. We can conclude from Figure 3.5 that the broadside array has the same average beampattern as the previous CR network. This is due to the result shown in equation (3.16). The $\alpha_b(\phi)$ in equation (3.18) can be approximated to $\alpha(\phi)$ defined in equation (3.15), when ϕ is close to ϕ_0. As a result in the angle area close to $\phi_0 = 0°$, the performances of the beampattern of the broadside array and the DB method are very similar to each other. In general the broadside array has a much narrower main beam than the end-fire array. In addition, to the same conclusion that has also been drawn in [18], we have discovered that the width of the main beam is in a reverse relationship with the size

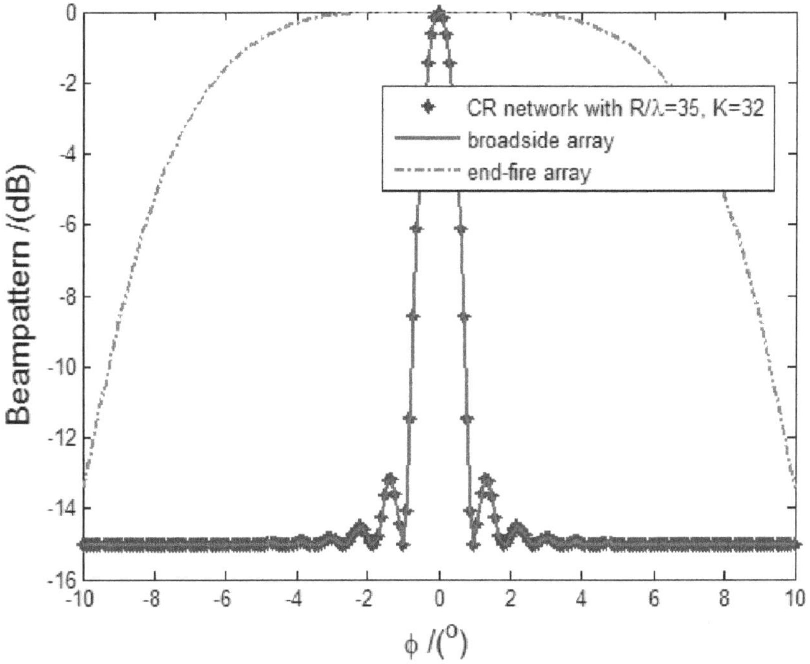

Figure 3.5 Beampattern of CR Network, Broadside Array and End-Fire Array

of the broadside array. Thus we are inspired by these two facts that if we want to enlarge the width of the main beam, we have to decrease the length (size) of the broadside array and we can adopt the end-fire array instead.

Based on this idea, we propose a NS method and select those nodes, which are able to form a full size end-fire array and a reduced size broadside array. Thus we choose the nodes in a relatively narrower belt along the DOA of the DCR user, as shown in Figure 3.6. In Figure 3.6, the CR nodes are selected in a way that the size of the "broadside" is limited to D, where $\frac{D}{\lambda} < \widetilde{R}$.

When we consider the case with more than one DCR user coexisting with the CR network, e.g. two DCR users, the NS method is demonstrated in Figure 3.7. We choose those CR nodes which are in the two "belt" areas as shown in Figure 3.7. In addition, for those double selected CR nodes, which are in the cross area, we adopt the method which let them randomly choose one of the two DCR users to serve, which has been proposed and discussed in [16].

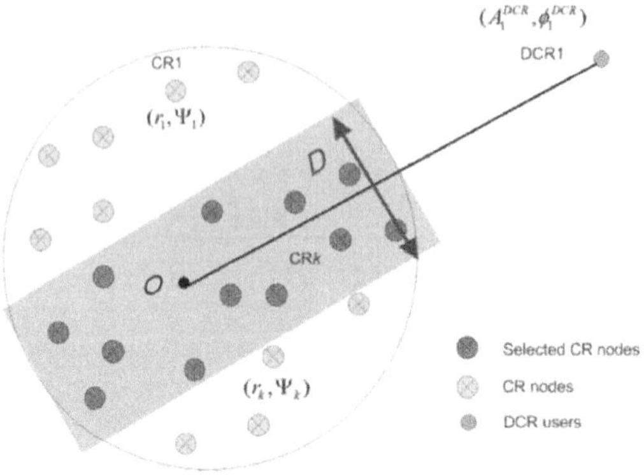

Figure 3.6 NS for CR Networks

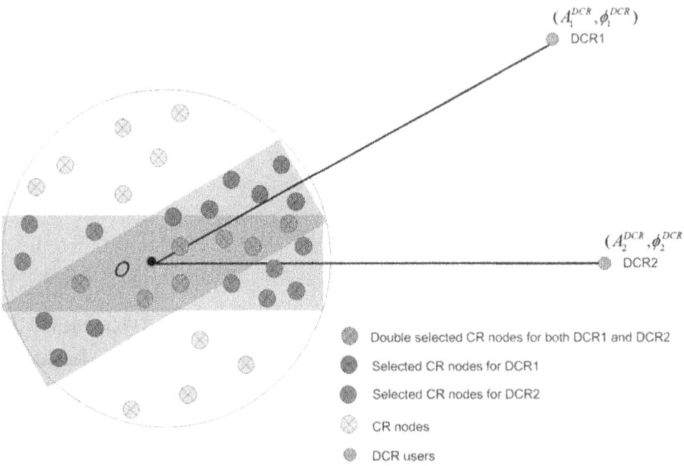

Figure 3.7 NS for CR networks with Two DCR Users

3.4 Simulation Results of the NS Method

Figure 3.8 and 3.9 demonstrate the selected nodes in the CR networks with different values of D by adopting the proposed NS method. Considering the fact that the number of selected nodes varies from each simulation, we average the number of selected nodes in order to show a general result of the beampattern. As a result, the beampattern of the NS method shown in Figure 3.10 is the average beampattern. In Figures 3.8–3.10, we assume there is only one DCR user and its DOA is $\phi_0 = 0°$. We choose the nodes within the width of the belt $D = 15\lambda$ and $D = 35\lambda$, as shown in Figure 3.8 and 3.9,

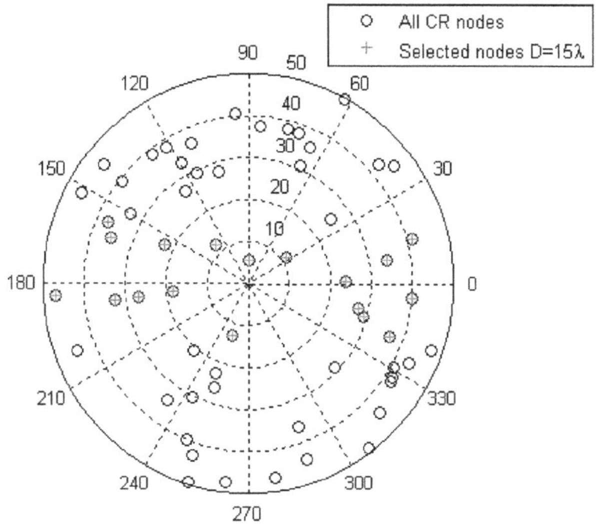

Figure 3.8 Selected CR Nodes in the CR Networks $D = 15\lambda(\phi_0 = 0°)$

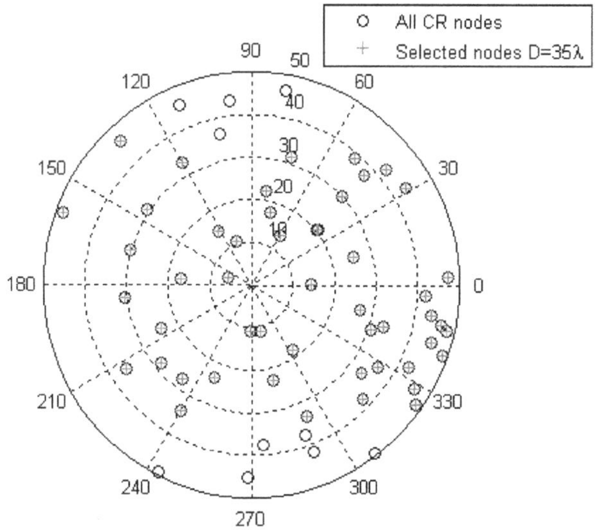

Figure 3.9 Selected CR Nodes in the CR Networks $D = 35\lambda$ ($\phi_0 = 0°$)

respectively. We assume there are 60 nodes in the network and the normalized radius of the network is 50, i.e., $K = 60$, and $\widetilde{R} = 50$. The result of the beampattern shown in Figure 3.10 is the average of 1000 runs.

Figure 3.8 and 3.9 show that those CR nodes, which are located in the belt area, as defined in Figure 3.6, are successfully selected for transmission. We can also see from these two figures that when *D* is smaller, less number of CR nodes will be selected for transmission.

Figure 3.10 shows that after adopting the proposed NS method, the main beam in the beampattern of the DB method is enlarged. Employing the NS method with defined $D = 15\lambda$ and $D = 35\lambda$, the main beam is about four times and two times wider than that without adopting the NS method, respectively. However, with smaller *D*, less CR nodes will be selected, as shown in Figure 3.6 and 3.8. Therefore the beampattern has a higher sidelobe level than that with a larger *D*, since the asymptotic sidelobe level of the beampattern is proportional to the reverse of the number of nodes of the CR network, as explained in Figure 3.3.

Figure 3.11 and 3.12 consider the case with two DCR users, which are located at $\phi_1 = 0°$and $\phi_2 = 30°$ when $D = 15\lambda$. Figure 3.11 shows the selected nodes in the CR networks to participate in CR transmission towards two DCR users. We can see in Figure 3.11 that there are a few CR nodes which are double selected for participating in transmission towards both DCR users.

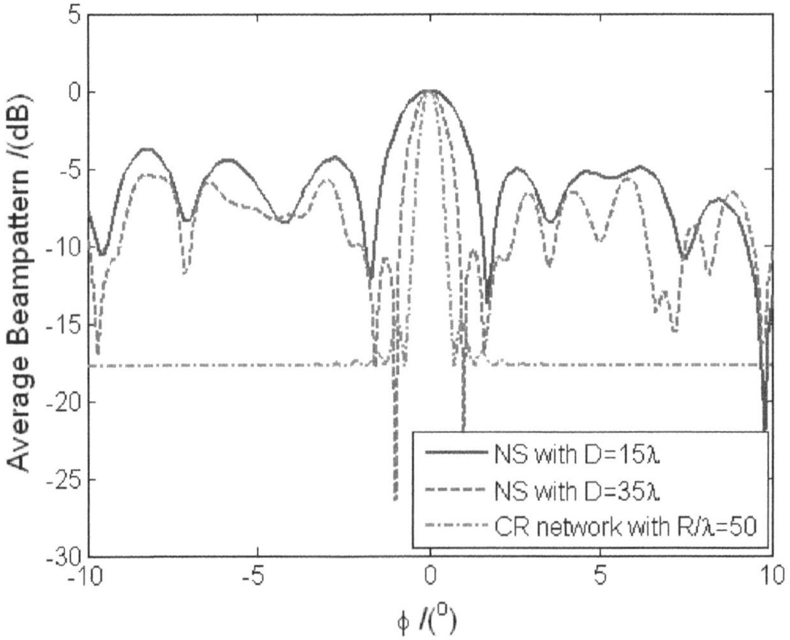

Figure 3.10 Average Beampattern of the Selected CR Nodes ($\phi_0 = 0°$)

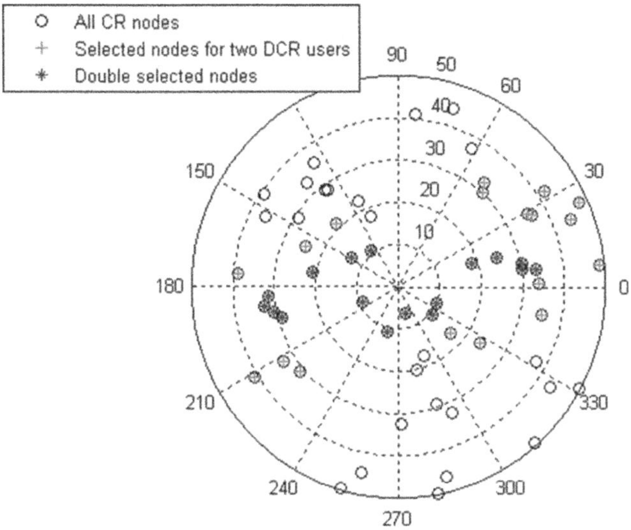

Figure 3.11 Selected CR Nodes in the CR networks $D = 15\lambda(\phi_1 = 0°$ and $\phi_2 = 15°)$

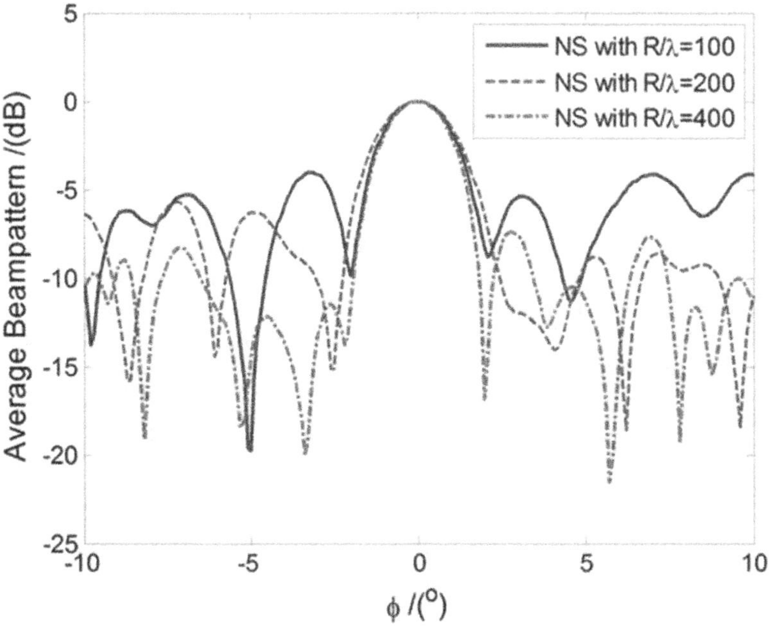

Figure 3.12 Average Beampattern of the Selected CR Nodes for Two DCR Users

Figure 3.12 shows the average beampattern of the NS method for two DCR users. As can be seen from this figure, the two main beams in the beampattern are directed towards $\phi_1 = 0°$ and $\phi_2 = 15°$, respectively. The main beams are both widened via adopting our NS method. When we employ the NS method with $D = 15\lambda$, we can see from Figure 3.12 that both two main beam are broadened to $3°$.

Figure 3.13 shows the result of the average beampattern of our proposed NS method when applied to large CR networks ($\widetilde{R} = 100, 200, 400$). We consider that the distribution density of the CR nodes remains the same with that of the Figure 3.8–3.12 ($K = 60$, $\widetilde{R} = 50$), and only the size of the networks is increased. Consequently the number of CR nodes of the considered CR networks in Figure 3.13 is $K = \frac{60\times100^2}{50^2} = 240, K = 960$ and $K = 3840$. We adopt $D = 15\lambda$ for all the three CR networks. It can be seen from this figure that the main beams are much wider than those of the DB method without NS method which are approximated by $\frac{35°}{\widetilde{R}}$. When the size of the CR network is enlarged, more nodes will be selected to participate in CR transmission. As a result sufficiently lower sidelobes can be achieved. For the case of $\widetilde{R} = 400$, far sidelobe levels become approximately $-15dB$, where near sidelobes are higher.

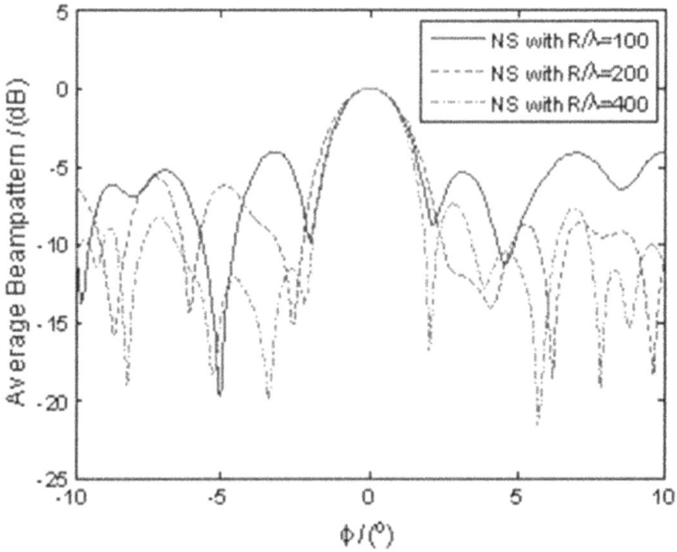

Figure 3.13 Average Beampattern of the Selected CR Nodes with $\widetilde{R} = 100, 200, 400$

3.5 Summary

We have introduced the DB technique to the CR network, which is constituted of distributed CR nodes. The goal of the DB method is to forward the CR signal to the DCR user, while causing no harmful interferences to coexisting PUs by limiting its transmission power towards directions PUs.

When introducing the DB method with the same original model of the CR network as shown in Figure 3.2, we notice that it is unavoidable that the original model will lead to the extremely narrow main beam of the beampattern. To find a new network structure which has less impact of the working frequency, we have proposed a Nodes Selection (NS) method. The NS method chooses those CR nodes, which are closely located to an artificial end-fire array considering the direction of the DCR user. Our NS method can also be applied to cases with more than one DR users. The average beampattern of the proposed NS method show that the main beams are successfully directed towards the DCR users and are sufficiently enlarged for practical applications in CR networks. What is more, for a CR network with a large physical size, our NS method can widen the main beam while maintaining adequate low sidelobe levels for CR transmission.

References

[1] Mitola III, J "Cognitive radio: an integrated agent architecture fro software defined radio," Ph.D., Royal institute of technology Stockholm, Sweden, 2002.

[2] Srinivasa, S. and S. A. Jafar, "The throughput potential of cognitive radio: a theoretical perspective," in Signals, Systems and Computers, 2006. ACSSC '06. Fortieth Asilomar Conference on, 2006, pp. 221–225.

[3] Akan, O., O. Karli, and O. Ergul, "Cognitive radio sensor networks," IEEE Network, vol. 23, pp. 34–40, 2009.

[4] Ochiai, H., P. Mitran, H. V. Poor, and V. Tarokh, "Collaborative beamforming for distributed wireless ad hoc sensor networks," IEEE Transactions on Signal Processing, vol. 53, pp. 4110–4124, 2005.

[5] Lun, D., A. P. Petropulu, and H. V. Poor, "A Cross-Layer Approach to Collaborative Beamforming for Wireless Ad Hoc Networks," IEEE Transactions on Signal Processing, vol. 56, pp. 2981–2993, 2008.

[6] Zarifi, K., S. Affes, and A. Ghrayeb, "Distributed beamforming for wireless sensor networks with random node location," in Acoustics,

Speech and Signal Processing, 2009. ICASSP 2009. IEEE International Conference on, 2009, pp. 2261–2264.

[7] Mudumbai, R., G. Barriac, and U. Madhow, "On the Feasibility of Distributed Beamforming in Wireless Networks," IEEE Transactions on Wireless Communications, vol. 6, pp. 1754–1763, 2007.

[8] Yindi, J. and H. Jafarkhani, "Network Beamforming Using Relays With Perfect Channel Information," IEEE Transactions on Information Theory, vol. 55, pp. 2499–2517, 2009.

[9] Gan, Z., W. Kai-Kit, A. Paulraj, and B. Ottersten, "Collaborative-Relay Beamforming With Perfect CSI: Optimum and Distributed Implementation," IEEE Signal Processing Letters, vol. 16, pp. 257–260, 2009.

[10] Havary-Nassab, V., S. Shahbazpanahi, A. Grami, and L. Zhi-Quan, "Distributed Beamforming for Relay Networks Based on Second-Order Statistics of the Channel State Information," IEEE Transactions on Signal Processing, vol. 56, pp. 4306–4316, 2008.

[11] Fazeli-Dehkordy, S., S. Shahbazpanahi, and S. Gazor, "Multiple Peer-to-Peer Communications Using a Network of Relays," IEEE Transactions on Signal Processing, vol. 57, pp. 3053–3062, 2009.

[12] Haihua, C., A. B. Gershman, and S. Shahbazpanahi, "Filter-and-Forward Distributed Beamforming in Relay Networks With Frequency Selective Fading," IEEE Transactions on Signal Processing, vol. 58, pp. 1251–1262, 2010.

[13] Brown, D. R. and H. V. Poor, "Time-Slotted Round-Trip Carrier Synchronization for Distributed Beamforming," IEEE Transactions on Signal Processing, vol. 56, pp. 5630–5643, 2008.

[14] Mudumbai, R., D. R. Brown, U. Madhow, and H. V. Poor, "Distributed transmit beamforming: challenges and recent progress," IEEE Communications Magazine, vol. 47, pp. 102–110, 2009.

[15] Zarifi, K., S. Affes, and A. Ghrayeb, "Collaborative Null-Steering Beamforming for Uniformly Distributed Wireless Sensor Networks," IEEE Transactions on Signal Processing, vol. 58, pp. 1889–1903, 2010.

[16] Lian, X., H. Nikookar, and L. P. Ligthart, "Efficient Radio Transmission with Adaptive and Distributed Beamforming for Intelligent WiMAX," Wireless Personal Commuications, vol. 59, pp. 405–431, 2011.

[17] Ahmed, M. F. A. and S. A. Vorobyov, "Collaborative beamforming for wireless sensor networks with Gaussian distributed sensor nodes," IEEE Transactions on Wireless Communications, vol. 8, pp. 638–643, 2009.

[18] Balanis, C. A., Antenna Theory: Analysis and Design, 3rd Edition: Wiley-Interscience, 2005.

Biography

Xiaohua Lian was born on May 10 1980 in Urmqi, P.R. China. She received her Bachelor and Master of Engineering respectively in June 2002 and April 2005 from Nanjing University of Aeronautics and Astronautics, China. In Oct. 2013 she received her PhD degree from Delft University of Technology, The Netherlands. Her areas of interest include Smart antennas, Beamforming and cognitive radio.

4

EEG Signal Processing for Post-Stroke Motor Rehabilitation

Silvano Pupolin[1], Giulia Cisotto[1] and Francesco Piccione[2]

[1]University of Padua, Department of Information Engineering, Padua, Italy
[2] IRCCS San Camillo Hospital Foundation, Department of Neurophysiology, Venice, Italy

4.1 Introduction

Ageing of population and modern life style standards in advanced countries are factors which favours the increase of stroke incidence in the population. Progress in the medical field is increasing the percentage of stroke survivors, but it is not able to guarantee to all of them an independent life, yet. Indeed, spontaneous recovery mostly occurs within the first six months then, when the patient enter the chronic state, several therapies can be employed to recover at least partially lost functions – especially motor functions [1].

The purpose of post-stroke rehabilitation consists in finding appropriate training to make patients become autonomous in their daily life activities by improving the ability to express properly their minds, recognize the ambient where they are and improve the mobility of their impaired limbs. In case of successful rehabilitation people could return to their normal activities while, if only partial recovery had occurred, they could be able to perform independently some basic activities at least. Even a partial success is a great improvement for patients, their families and even societies. In fact, if a smaller amount of people has to be assisted with intense and expensive 24-hours care, societies can fund money to other projects.

Hereafter we limit our analysis to motor rehabilitation procedures to improve patient performance in activities like reaching or grasping.

Convergence of Communications, Navigation, Sensing and Services, 71–90.

In standard rehabilitation programs the patient usually undergoes physiotherapy to re-establish motor functionalities compromised by stroke consequences. Improvements are due to the *brain plasticity*, i.e. the ability of the brain to exploit new neural paths to perform functions previously controlled by other stroke-injured neurons.

A recent approach in motor rehabilitation of the upper limb includes the use of a robotic device which moves the damaged arm following a prescribed set of movements within a training period of several weeks. During the exercise the robot measures all the forces involved in performing the task looking at the patient reactivity and applies a feedback related to that. Literature reports improvements in patients treated with this kind of approach. It can be noted that in this experiment no measures of the brain activity are provided, thus it appears as a robotic repetitive version of the standard physiotherapy procedure with a better performance measurement [2].

However, neurologists believe that patients could reach a better motor performance if some re-learning strategy is adopted to force the patient to activate a new healthy neural path, closed to the damaged ones, to control the movement (*neural plasticity*). In recent years *Brain Computer Interface* (BCI) [3] has been exactly proposed with this aim and, specifically, as a tool to improve motor rehabilitation by exploiting neural plasticity [4]. Proprioceptive feedback seems to be a critical element in the rehabilitation program [5]: several studies have demonstrated that BCI training with this kind of feedback and a concurrent physical therapy activity can highly improve the rehabilitation process [6, 7].

Combining the robotic intervention and the exploitation of the neural resources together, a sophisticated rehabilitation approach has been recently proposed. It is based on a robot assistance of the movement of the affected arm of the patient based on the identification of the neuronal populations involved in the task execution. Indeed, a real-time feedback to the movement is given on the basis of the area where those neuronal populations are activated. Real-time is the most stringent constraint in designing such a system, because it means to give a feedback in a fraction of second. In order to identify the area where the activated neurons lay we could use current diagnostic systems as *functional Magnetic Resonance Imaging* (fMRI), but this tool requires several seconds to process the incoming signals, therefore it is not suitable for our purposes. Another classical diagnostic device is *ElectroEncephaloGram* (EEG) which combined with an appropriate signal processing gives a *low resolution tomography* (LORETA). By using LORETA we can approximately localize the activated neurons within the strict real-time constraint required in the application we are considering. Then, an EEG-based BCI will be presented in this chapter along with some preliminary results.

The Chapter is organized as follows. Section 2 is devoted to identify the EEG signals of interest for the motor rehabilitation, while Section 3 shows in details the mechanism behind the so called *operant-learning rehabilitation*, which is a key feature for this new rehabilitation procedure. Section 4 deals with EEG signal processing and the constraints imposed by the real-time requirements. Section 5 reports an example of application of the above real-time signal processing method with some preliminary results. Section 6 concludes the Chapter proposing further research activities.

4.2 Neurophysiological Signal Analysis for Motor-Rehabilitation

When we measure the electrical activity of the brain by an EEG or a *Magneto-EncephaloGram* (MEG) we note the presence of rhythmic activities in which frequencies depend on the state of the person and on the activities he/she is performing. It has been noted that several kinds of events can induce time-locked changes in the activities of neuronal populations that are called *event related potentials* (ERP). On the other side, phase-locked spontaneous oscillations can be stimulated in specific neural populations with the subject performing other types of tasks. The latter, named *event-related de-synchronization* (ERD), or *event-related synchronization* (ERS), can be measured by means of an EEG as a power decrease or increase, respectively, of specific spectral components.

However we remark that the brain is organized in such a way to operate in a multitasking scheme, i.e. several activities can be accomplished simultaneously using different resources. Thus, the signal measured by one EEG sensor results as a linear combination of all the signals generated by the cerebral inner sources. Moreover a specific frequency band is dedicated to each of the above mentioned activities, with the exact frequency limits varying from person to person and on the basis of the instantaneous physical state, emotional and stress conditions.

Then, the analysis of ERD/ERS is quite difficult and should be tailored to the specific activity we want to analyze and to the state of the person. Moreover, if we know the cerebrum volume where this activity takes place, we could spatially separate the sources of the signals and take care only of the ones we are interesting in.

To this purpose, we recall that the brain is divided in two hemispheres (left and right) and each of them is further split into four lobes (frontal, temporal, parietal, and occipital). They are separated by the longitudinal fissure and connected via the corpus callosum, the anterior and the posterior commissures. The frontal lobe is separated from the parietal lobe by the central sulcus and from the temporal lobe by the lateral fissure. The parieto-occipital sulcus divides the parietal from the occipital lobe. The cerebral cortex, composed mainly by neurons, covers the whole hemispheres with a thin layer (2–3 mm thick). It is typically divided into three functional areas: i) *the sensory areas* which receive and interpret somatic sensory impulses including cutaneous sensations, input from the five senses and some aspects of proprioception, ii) *the association areas* which integrate sensory information with emotional states, memories, learning and rational thought processes and iii) *the motor areas* which generate impulses which innervate voluntary skeletal muscles.

All the activities performed by the functional areas induce the generation of electrical signals that could be measured by an EEG. As mentioned above, many activities could be run simultaneously so that many electrical signals arise from distinct areas and combine together at the EEG electrodes. Moreover it has been recognized that during the idle state some areas generate a periodic signal (wave or rhythm) that results from a synchronous activity of many neurons of the same functional area.

Hereafter we limit the analysis to the cerebral signals that are related to movement and its preparation. As a hypothesis, we can state that all the waves we analyse are due to synchronous and coherent electrical activity of neural cells. The waves we are interested in are illustrated hereafter.

Alpha waves. They belong to the frequency range of 8–13 Hz. We limit our analysis to the Alpha waves during relaxed mental state, where the subject is at rest with eyes closed, but not tired or asleep. This Alpha activity is centered in the occipital lobe and is presumed to originate there, although there has been recent speculation that it has instead a thalamic origin.

Beta waves. They have a frequency range of 12–30 Hz. Low amplitude Beta waves with multiple and varying frequencies are often associated with active, busy, or anxious thinking and active concentration conditions.

Over the motor cortex Beta waves are associated with the muscle contractions occurring during isotonic movements and are suppressed before to and during movements [8]. Bursts of Beta activity are associated with a strengthening of sensory feedback in static motor control and its amplitude increases after a movement or when an action against an opposition has to be performed [9].

Mu waves. Also known as sensorimotor or Rolandic rhythms, they are due to neural activity in the motor cortex [10]. These patterns as measured by an EEG or a MEG have a frequency range of 8–13 Hz and are most prominent when the body is physically relaxed [10]. Mu waves are suppressed when the person performs a motor action or even when he/she observes the same action performed by someone else. Although no general consensus is expressed by the medical community, some researchers have interpreted the latter phenomenon as the involvement of the mirror neurons in the Mu wave suppression [11, 12, and 13].

Mirror neurons have been studied since 1990 in macaque monkeys (using invasive measuring techniques) and later in humans (using EEG and fMRI). The study has shown that mirror neurons fire during basic motor tasks and also when monkeys and humans observe others performing the same simple tasks [15]. Furthermore it appears that mirror neurons have components that deal with intention [19]. They are located in the right fusiform gyrus, left inferior parietal lobe, right anterior parietal cortex, and left inferior frontal gyrus [14, 17, 18]. Mu waves are suppressed when mirror neurons fire. This suppression should represent a higher-level coordination of mirror neuronal activity throughout the brain [12, 13, and 16].

Suppression of Mu waves is also called *desynchronization* and it usually occurs during voluntary movements. Generally speaking, this kind of task produces a wider desynchronization: not only Mu rhythms, but also upper Alpha and lower Beta bands are involved, indeed. This phenomenon starts about 2s before the movement-onset over the contralateral Rolandic region and becomes bilaterally symmetrical immediately before its execution.

ErrP waves: Error Potential (ErrP) waves have two main components in an error trial: a negative potential (Ne) showing up as a sharp negative component peaking at about 100 ms after the incorrect movement, and a positive potential (Pe) showing up as a broader positive component with a peak between 200 and 500 ms after the incorrect movement. More generally, ErrPs are associated with error processing. Literature showed the presence of the Ne even in correct trials and proposes that Ne reflects a comparison process, in both correct and erroneous trials. Pe is a further error-specific component, independently of Ne, and it is associated with a later aspect of error processing [20]. Other studies focused on awareness of errors: they showed that irrespective of whether the subject is aware of the error or not, erroneous trials are followed by Ne. In contrast Pe is much more pronounced for perceived errors. These findings verify the hypothesis that Ne reflects an unconscious comparison process and Pe reflects a later conscious error processing. This kind of signal is generated in the anterior cingulate cortex that is a deep fronto-central cortical area. The best EEG pins to detect ErrP wave are Cz and Fz, indeed.

All the above signals are useful for the BCI upper limb training that will be shown in Section 3

4.3 Neuroplasticity Enhancement and an Operant-Learning Protocol

Neuroplasticity is the key issue for rehabilitation after a stroke. Actual clinical solutions pursue innate physiological and anatomical plasticity processes that underlie substantial gains in motor function after a stroke [21, 22] by means of the combination of task specific training with general aerobic exercises. However 15–30% of patients with a stroke remain with permanent disabilities. Novel therapies to enhance neuroplasticity would be thus highly welcome. To summarize, current researches focus on the following four main research areas:

1. The first relates to the study of the molecular and cellular mechanisms of normal movement and the pathophysiological processes involved in post stroke paresis. An in depth understanding of these mechanisms should lead to improvements in prognostic indicators of functional recovery and more effective interventions to improve relearning of lost motor functions in comparison with those currently available [23–25].
2. The second area is concerned with the development of pharmacological, biological and electrophysiological techniques that can increase training-induced plasticity [26].

   Both of these research areas aim to understand and enhance innate plastic mechanism in the Central Nervous System (CNS), so that these neural processes can be harnessed to aid post stroke rehabilitation [21, 27].
3. The third area aims to utilize developments in biomedical and tissue engineering to promote functional recovery within the brain. It has to be mentioned that preliminary attempts to promote neural repair through use of innate or exogenous stem cells have also been made [28, 29].
4. The fourth area involves the development of neuroprosthestics and BCI technologies to bypass injury via adaptive remote neuroplasticity, whereby part of the nervous system not originally dedicated to a particular task can be harnessed to provide the neural substrate that interacts with the neuro-prosthesis or brain–computer device [30].

Hereafter we describe an experiment we are conducting based on the fourth research area. We should solve several clinical and technical problems so that the BCI system could reach the target results, i.e. to reduce the percentage of stroke patients which remain permanently disabled.

In order to introduce an operant learning-based rehabilitation we need to implement a closed loop in which penalizations and rewards for the execution of the test are present. The subject was asked to perform a standard center-out reaching task on a horizontal plane, grasping the *end-effector* of a robotic device. The task was considered properly completed if the patient reached, along a 18 cm-path, the target – one out of four cardinal points, randomly chosen by the software – within a time window between 500 and 740 ms. The system included an EEG which captured cerebral signals, processed them in order to understand if the expected part of the brain has been activated to perform the experiment. Otherwise, the penalization consisted in a force feedback opposing to the movement of the end effector.

This implemented an operant-learning scheme for the rehabilitation.

Specifically, the protocol we designed should be able to close the stroke-impaired sensorimotor loop of the subject by applying the proper force to the end effector of the robot when the upper limb movement begins; taking also into account the delays occurred during the signal processing.

The patient seated in front of a screen where the position of the end-effector of the robotic arm he/she grasped was displayed. An auditory signal was activated at the beginning of specific time intervals that scanned each trial. Particularly, after an initial rest period of about 40 s the patient performed several trials. Each of them was composed by the following intervals:

- *Pre-trigger*, an initial window of 500 ms during which the subject was at rest;
- *Post-trigger*, beginning after the appearance on the screen of the target cardinal point to reach, it lasted for 1500 ms; the subject had to wait in the starting position until an auditory signal was delivered.
- *Reaction-time*, the time to actually move from the starting position after the auditory signal that allowed the subject to move. The duration of this phase individually varied but generally lied in the (400,700) ms range;
- *Movement*, the time interval to move from the starting position to the final destination. It was considered correctly performed if the subject completed it within the time interval between 500 and 740 ms. As the Reaction time, it considerably varied from one subject to another.

- *Return*, after completing the movement, the patient was asked to return to the initial position and relax before starting the next trial.

As mentioned before, the protocol was designed by taking into account the physiology of the waves we measured by the EEG and in particular the timing of the sensorimotor rhythms desynchronization generally starting 2 seconds before the movement.

It is worth to mention that even the latter is a variable value: therefore, a minimum value of 1.9s was considered. This value represented then the most critical parameter of the whole system since many signal processing operations have to be performed within this lapse. Specifically, the following actions were required:

1. Filter the incoming EEG signals;
2. Process them to capture the information about the frequency and spatial distributions of the de-synchronization occurring during the trial;
3. Decide if the de-synchronization occurred in the expected area of the brain or not;
4. Apply the related force to the end-effector on the basis of the results at step 3).

Then, in the next Section we analyze in detail the design of the filters and the signal processing algorithms in order to provide a suitable force feedback together with fulfilling the strict time requirements.

4.4 Constraints on Signal Processing to Implement an Operant-Learning Protocol

Operant learning rehabilitation is effective if we are able to give feedback to the person in real time. The meaning of real time is in some sense misleading and need to be explained in details. Operant learning rehabilitation is based on a closed loop BCI system. In our case, where we are looking for upper limb rehabilitation, the process is based on a protocol to perform non repetitive exercises. The purpose of the rehabilitation is twofold: a) to improve the capacity of the patient to move the affected upper limb, and b) to use part of the brain closed to the one damaged to do the movement. The protocol is designed in such a way that we are able to identify by EEG the part of the brain involved in the movement by measuring the de-synchronization of Alpha, Beta and Mu waves. These phenomena are more relevant in different part of the scalp. Moreover, to obtain meaningful parameters related to this de-synchronization it is appropriated to filter the EEG signals within the specific bands in which the phenomena take place. Filtering allows reducing noise and interference from other brain signals, so that the de-synchronization is more visible. As mentioned in Section 2 it is well-known that de-synchronization of Mu waves occurs when imaging subject images to perform a movement. It appears on the contralateral hemisphere with respect to the active limb, while later on it shifts to a more central part of the brain.

In order to separate out the frequency components we filter the incoming EEG signals with bandpass filters centered on 10.5 Hz for Alpha and Mu waves and on 21 Hz for Beta waves. The filter bandwidth is 5 Hz for Alpha and Mu waves and of 18 Hz for Beta waves. Moreover, we recall that a real band-limited filter causes the output to be delayed by a time interval that can be estimated to be twice the inverse of the bandwidth; therefore, we expect a delay of about 400 ms for Alpha and Mu waves and of about 100 ms for Beta waves. Then the protocol we design should take into account these delays in order to make the end effector of the robot ready to properly act when the actual movement begins.

The EEG signal is sampled at 512 Hz with samples consequently spaced by 2ms each other. Therefore, regarding the required signal processing actions explained at the end of Section 3, it has to be noticed filtering step i) causes a delay of about 400 ms; at step ii) 256 signal samples are analysed at any time adding a further delay of 500 ms about. Finally, based on the designed protocol, 1 s lasts to perform tasks iii) and iv). This time is sufficiently long and allows us to perform further signal analysis in order to get a more significant quantification of the de-synchronization phenomenon.

4.5 Preliminary Results

The protocol described in the previous sections was, indeed, implemented at I.R.C.C.S. San Camillo Hospital Foundation at Lido of Venice (Italy) – a rehabilitation Institute for patients suffering from neurological diseases [31, 32 and 33]. The voluntary modulation of *sensory motor rhythms* (SMR), especially the Mu and Beta ones, was exploited to provide the BCI control. Specifically, as already suggested, this BCI platform allowed stroke survivors to control a robotic arm that helped them in accomplishing a standard center-out reaching exercise on a plane.

Actually, the system setup is captured by Figure 4.1. The rehabilitation program lasted for about three weeks and was constituted by three main phases:

1) *Screening session*, in which the subject was clinically evaluated by means of a battery of clinical tests and a preliminary BCI session was performed.

Figure 4.1 Experimental Setup

At the end of this first day, a set of EEG features could be identified and used to control the BCI during the subsequent training period.

2) *Training sessions:* this was the actual BCI treatment. It approximately lasted two weeks and during this period of time the patient was required to perform eighteen BCI sessions repeating – following the strict experimental protocol described in Section 3 – the reaching movement towards the target shown on the screen in front of the subject. As aforementioned in Section 3, in order to benefit of the robotic help to conclude the trial, the patient had to modulate his/her Mu or lower Beta rhythms accordingly to the movement or the rest condition: therefore, he/she had to enhance the amplitude of those EEG signals during rest and lower them during the planning (post-trigger time) and the actual execution (movement and return) of the movement. On the basis of the quite lacking literature on this specific topic, we hypothesized that the particular kind of proprioceptive feedback employed in this experiment should improve and fasten the ability of the patient to voluntarily – although non-fully consciously – modulate their SMR rhythms. This consequently should lead to a reinforcement of the training efficiency with a following longer-lasting benefit even after the hospitalization period.
3) *End-test session:* this took place on the final day of the experimental program in which a last evaluation was administered to the subject. It was fully-identical to the screening one and it provided a set of clinical,

kinematic and neurophysiological measures to evaluate the improvement of the subject after the treatment and the system efficiency.

In a very first phase, a healthy person was actually recruited to test the system. He performed only the screening session and his kinematic outcomes are reported in Table 4.1.

In the following, one right-sided capsular ischemic stroke patient with a mild impairment on his left arm was involved in the study. Because of the lack of a larger experimental population, only preliminary -although promising- results could be reported here (see Table 4.2).

Measurements provided by Table 4.2 display an average improvement in the kinematic of the patient's left-affected arm. In fact, to summarize these results, at the end of the training period he scored a higher number of correct trials (out of a total of 240) performing more and more accurate and faster movements: the general kinematic improvement can be shown by the decrease of the *area error* – the squared distance between the straight ideal path and the actual trajectory from the starting position to the target - the decrement of the movement duration and the related increment of the motion speed.

For the sake of completeness Table 4.3 reports the performance of the right healthy arm, useful as a comparison with the damaged one.

Table 4.1 Kinematic Performance of a Healthy Subject During the Screening Session

	% Correct trials	Duration (average, standard dev.) [ms]	Area Error (average, standard dev.) [mm^2]
Right arm	85.9	(615, 89)	(12.3, 22.3)
Left arm	76.6	(647, 104)	(23.5, 32.5)

Table 4.2 Kinematic Performance with the Left-Affected Arm

	% Correct trials	Duration (average, standard dev.) [ms]	Area Error (average, standard dev.) [mm^2]
Screening	24.2	(902, 244)	(35.2, 46.8)
End-test	44.2	(811, 209)	(32.5, 36.5)

Table 4.3 Kinematic Performance with the Healthy Right Arm

	% Correct trials	Duration (average, standard dev.) [ms]	Area Error (average, standard dev.) [mm^2]
Screening	50.0	(781, 216)	(23.2, 28.9)
End-test	61.3	(729, 136)	(22.4, 25.5)

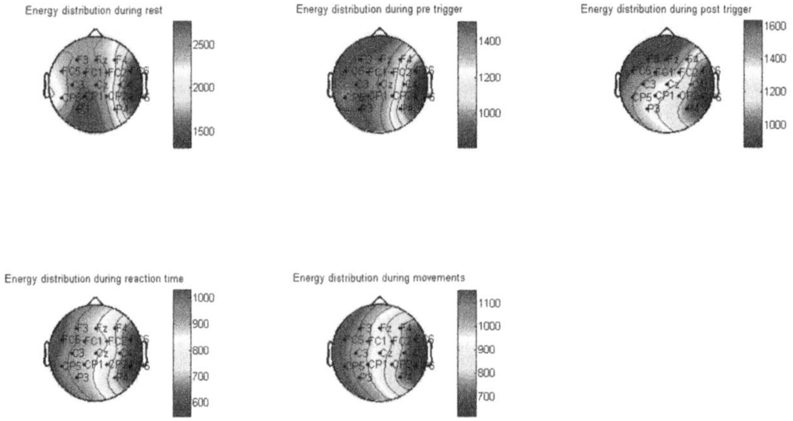

Figure 4.2 Mean Energy Distribution in (10, 14) Hz Band During Rest and in each Phase of the Trials for the right Healthy Arm Movements

As the EEG signals analysis regards, the presence of the movement-related de-synchronization was assessed by a power spectral analysis on the (10, 14) Hz band as shown, for example, in Figures 4.2 and 4.3.

The latter topographical maps represent the mean spectral energy of the EEG signals in the (10, 14) Hz band at the various locations over the scalp during the initial relaxation period at the beginning of each experimental run and in the four phases (pre-trigger time, post-trigger time, reaction time and movement) of each of the following 80 trials. An example of the spectral power decrease along these phases is provided by Figures 4.2 and 4.3 in the case of healthy right-arm movements and damaged left-arm ones, respectively.

Looking at Figure 4.2 we remark that from rest to pre-trigger there is an energy reduction of about 60% and a further reduction of about 70% from post-trigger to reaction time occurs, while during the movement the measured energy begins to rise-up again. Moreover, we can observe that there is a higher energy in the left lobe than in the right one in all phases in agreement with the contra-lateral activation of the brain with respect to the moving arm. On the other hand, the high energy values of the frontal areas of the brain during post-trigger could be associated with the visualization of the target to reach and with the associated cognitive brain processes.

The so-called *explained variance* or *coefficient of determination* (R^2) can be also estimated and is commonly used to integrate the analysis of the *Movement Related De-synchronization* (MRD) phenomenon. R^2 is a standardized statistical quantity that infers the difference between two distributions, i.e. that

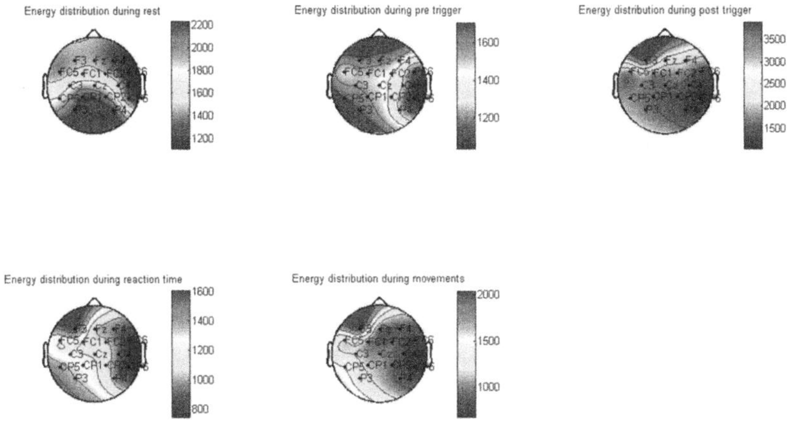

Figure 4.3 Mean Energy Distribution in (10, 14) Hz Band During Rest and in Each Phase of the Trials for the Left Affected Arm Movements

of the relax period and that of a specific phase of the trials for example, taking into account their variances also.

As far as high R^2 values represent large differences between the two conditions, the higher the R^2, the larger the MRD phenomenon. An example of R^2 distribution is displayed in Figures 4.4 and 4.5 and it can be noted that the maximum evidence of the MRD phenomenon is focused over the C4 and CP2 electrodes in the contralateral hemisphere – the ipsilesional one – to the

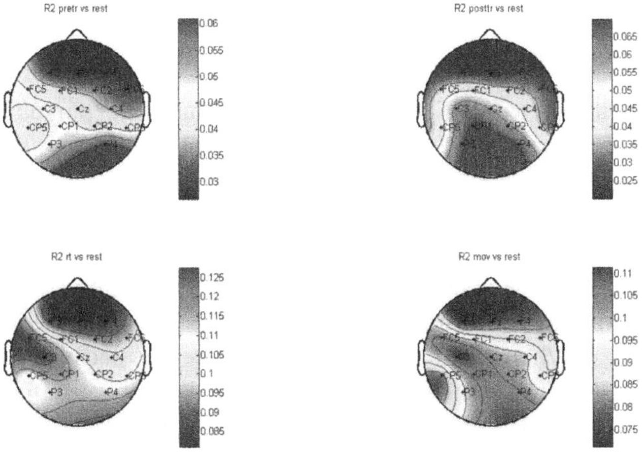

Figure 4.4 R^2 Distribution in (10, 14) Hz Band in each Phase of the Trials for the Right Healthy Arm

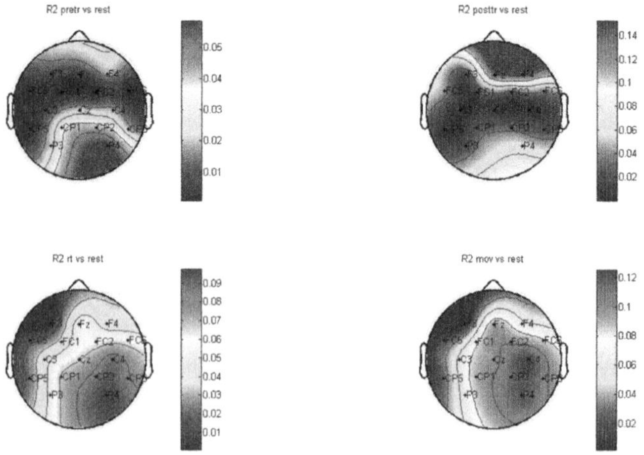

Figure 4.5 R^2 Distribution in (10, 14) Hz Band in each Phase of the Trials for the Left-Affected Arm

movement and in the C3 and CP1 electrodes for the right arm movement, respectively, as expected from the neurophysiological literature.

Further analysis are currently being carried on in order to better characterize the neurophysiological patterns related to the movement and, consequently, customize the treatment on the basis of the most significant EEG features of the patient even adapting the feedback to his/her improvements during the training.

4.6 Conclusions and Future Goals

Preliminary results shown in Section 5 are encouraging to pursue a clinical systematic and controlled study in order to evaluate the improvement of patients from his/her disability by using this new BCI rehabilitation protocol.

During the clinical tests we could extend the data set already collected analyzing not only the de-synchronization but also other kinds of patterns, such as ErrPs, in the EEG signals and check the consciousness of the patient regarding his/her mistakes during the exercises and compute the correlation of the latter signals with EEG. Moreover the acquisition of electromyography signals from the upper limb could be correlated to the EEG signals to find if the actual neurophysiological patterns of the intention to move.

Furthermore, the use of a robot to perform upper limb movements as proposed by Piovesan et. al. [2] with a closed loop BCI could be helpful for the rehabilitation of more severe impaired patients.

The efficacy of this treatment will then be measured on the basis of the treatment duration and its benefits for the patient. A comprehensive statistical evaluation based on a larger population is also required. To this purpose, a long term strategy has to be planned and the results will likely be available in a couple of years.

Finally, a more precise clinical assessment score than the standard ones, as e.g.: Functional Independence Measure (FIM), Fugl-Meyer Assessment for the Upper Extremity (FMA-UE), Ashworth Modified Scale (MAS), Nine Hole Peg Test (NHPT), Box and Blocks Test (BBT) and Reaching Score (RS), will be proposed in order to capture even slighter improvements. In our opinion, this could be accomplished by gathering a more detailed quantification of the patient performance during the reaching task by means of the BCI system presented in this chapter.

References

[1] Piron L., Turolla A., Agostini M., Zucconi C.S., Ventura L., Tonin P., Dam M., "Motor learning principles for rehabilitation: A pilot randomized controlled study in post stroke patients", Neurorehabil Neural Repair, vol. 24, pp. 501–508, 2010.

[2] Piovesan D., Morasso P., Giannoni P., Casadio M., "Arm stiffness during assisted movement after stroke: the influence of visual feedback and training", IEEE Trans. Neural Systems and Rehab. Eng., vol. 21, pp. 454–465, 2013.

[3] Birbaumer N., Cohen L.G., "Brain-computer interfaces: communication and restoration of movement paralysis", J Physiol, vol. 579, pp. 621–636, 2007.

[4] Dimyan M.A., Cohen L.G., "Neuroplasticity in the context of motor rehabilitation after stroke", Nat. Rev. Neurol., vol. 7, pp. 76–85, 2011.

[5] Ramos-Murguialday A., Halder S., Birbaumer N., "Proprioceptive feedback in BCI", Proc of NER'09 4th Int. IEEE EMBS Conf. On Neural Engineering, Antalya, Turkey, pp. 279–282, 2009.

[6] Silvoni S., Ramos-Murguialday A., Cavinato M., Volpato C., Cisotto G., Turolla A., Piccione F., Birbaumer N., "Brain-Computer interface in stroke: A review of progress", Clin. EEG Neurosci., vol. 42, pp. 245–252, 2011.

[7] Broetz D., Braun C., Weber C., Soekadar S.R., Caria A., Birbaumer N., "Combination of brain-computer interface training and goal directed physical therapy in chronic state: A case report", Neurorehabil. Neural Repair, vol. 24, pp. 674–679, 2010.

[8] Baker, S.N., "Oscillatory interactions between sensorimotor cortex and the periphery". Current opinion in neurobiology, vol.**17**, pp. 649–55, 2007.

[9] Lalo, E., Gilbertson, T., Doyle, L., Di Lazzaro, V., Cioni, B., Brown, P., "Phasic increases in cortical beta activity are associated with alterations in sensory processing in the human". Experimental brain research. Experimentelle Hirnforschung. Experimentation cerebrale, vol. **177,** pp. 137–45, 2007.

[10] Florin, A., Lopes da Silva, F., "Cellular Substrates of Brain Rhythms". In Schomer, D. L.; Lopes da Silva, F. Niedermeyer's Electroencephalography: Basic Principles, Clinical Applications, and Related Fields (6th Ed.). Philadelphia, Pa.: Lippincott Williams & Wilkins. pp. 33–63, 2010.

[11] Oberman, L.M., Hubbarda, E.M., Altschulera, E.L., Ramachandran, V.S., Pineda, J.A., "EEG evidence for mirror neuron dysfunction in autism spectrum disorders". Cognitive Brain Research, vol. 24, pp. 190–198, 2005.

[12] Pineda, J.A., "The functional significance of mu rhythms: Translating "seeing " and "hearing "into "doing". Brain Research Reviews, vol. 50, pp. 57–68, 2005.

[13] Churchland, P., Braintrust: What Neuroscience Tells Us About Morality. Princeton, NJ: Princeton University Press. p.156, 2011.

[14] Williams, J.H.G., Waiter, G.D., Gilchrist, A., Perrett, D.I., Murray, A.D., Whiten, A., "Neural mechanisms of imitation and 'mirror neuron' functioning in autistic spectrum disorder". Neuropsychologia, vol. 44, pp. 610–621, 2006.

[15] di Pellegrino, G., Fadiga, L., Fogassi, L., Gallese, F., Rizzolatti, G., "Understanding motor events: A neurophysiological study". Experimental Brain Research, vol. 91, pp. 176–180, 1992.

[16] Rizzolatti, G., Fogassi, L., Gallese, V., "Neurophysiological mechanisms underlying the understanding and imitation of action". Nature reviews. Neuroscience, vol. 2, pp. 661–670, 2001.

[17] Marshall, P.J., Meltzoff, A.N., "Neural mirroring systems: Exploring the EEG mu rhythm in human infancy". Developmental Cognitive Neuroscience, vol. 1, pp. 110–123, 2011.

[18] Keuken, M.C., Hardie, A., Dorn, B.T., Dev, S., Paulus, M.P., Jonas, K.J., Den Wildenberg, W.P., Pineda, J.A., "The role of the left inferior frontal gyrus in social perception: an rTMS study". Brain Research, vol. 1383, pp. 196–205, 2006.

[19] Sinigaglia, C., Rizzolatti, G., "Through the looking glass: self and others". Consciousness and cognition, vol. 20, pp. 64–74, 2011.

[20] Vidal, F., Hasbroucq, T., Grapperon, J., Bonnet, M., "Is the "error negativity" specific to errors?" Biological Psychology, vol. 51, pp.109–128, 2000.

[21] Nudo, R.J., Wise, B.M., SiFuentes, F., Milliken, G.W., "Neural substrates for the effects of rehabilitative training on motor recovery after ischemic infarct". Science, vol. 272, pp. 1791–1794, 1996.

[22] Taub, E., Uswatte, G., Elbert, T., "New treatments in neuro-rehabilitation founded on basic research". Nat. Rev. Neurosci., vol. 3, pp. 228–236, 2002.

[23] Krakauer, J.W., "Motor learning: its relevance to stroke recovery and neurorehabilitation". Curr. Opin. Neurol., vol. 19, pp. 84–90, 2006.

[24] Seitz, R.J., "How imaging will guide rehabilitation".Curr. Opin. Neurol., vol. 23, pp. 79–86, 2010.

[25] Dimyan, M.A., Cohen, L.G., "Contribution of transcranial magnetic stimulation to the understanding of functional recovery mechanisms after stroke". Neurorehabil. Neural Repair, vol. 24, pp. 125–135, 2010.

[26] Floel, A., Cohen, L.G., "Recovery of function in humans: cortical stimulation and pharmacological treatments after stroke". Neurobiol. Dis., vol. 37, pp. 243–251, 2010.

[27] Buonomano, D.V., Merzenich, M.M., "Cortical plasticity: from synapses to maps". Annu. Rev. Neurosci., vol. 21, pp. 149–186, 1998.

[28] Lindvall, O., Kokaia, Z., "Stem cells for the treatment of neurological disorders". Nature, vol. 441, pp. 1094–1096, 2006.

[29] Delcroix, G.J., Schiller, P.C., Benoit, J.P., Montero-Menei, C.N., "Adult cell therapy for brain neuronal damages and the role of tissue engineering". Biomaterials, vol. 31, pp. 2105–2120, 2010.

[30] Wolpaw, J.R., Birbaumer, N., McFarland, D.J., Pfurtscheller, G., Vaughan, T.M., "Brain–computer interfaces for communication and control". Clin. Neurophysiol., vol. 113, pp. 767–791, 2002.

[31] Cisotto, G., Pupolin, S., Silvoni, S., Cavinato, M., Agostini, M., Piccione, F., "Brain-Computer Interface in Chronic Stroke: an application of sensorimotor closed-loop and contingent force feedback", Proc. IEEE ICC'2013, Budapest, Hungary, June 9–13, 2013

[32] Cisotto, G., Pupolin, S., Silvoni, S., Piccione, F., "An application of Brain Computer Interface in chronic stroke to improve arm reaching function exploiting Operant Learning strategy and Brain Plasticity", Proc. IEEE HEALTHCOM 2013, Lisboa, Portugal, October 9–12, 2013.

[32] Cisotto, G., Pupolin, S., Cavinato, Piccione, F., "An EEG-based BCI platform to improve arm reaching ability of chronic stroke patients by means of an operant learning training with a contingent force feedback", International J. of e-health and medical communications, to be published 2014.

Biographies

Silvano Pupolin (S67, M71, SM83), received the Laurea degree in Electronic Engineering from the University of Padova, Italy, in 1970. Since then he joined the Department of Information Engineering, University of Padua, where currently is Full Professor of Electrical Communications. He was Chairman of the Faculty of Electronic Engineering (1990-1994), Chairman of the PhD Course in Electronics and Telecommunications Engineering (1991–1997), (2003–2004) and Director of the PhD School in Information Engineering (2004–2007). Chairman of the board of PhD School Directors of the University of Padua (2005–2007), Member of the programming and development committee of the University of Padua (1997–2002), Member of Scientific Committee of the University of Padua (1996–2001), Member of the budget Committee of the Faculty of Engineering of the University of Padua (2003–2009), Member of the Board of Governor of CNIT "I National Interuniversity Consortium for Telecommunications" (1996–1999), (2004–2007), Director of CNIT (2008–2010), General Chair of the 9-th, 10-th and 18-th Tyrrhenian International Workshop on Digital Communications devoted to "I Wireless Communications", "I Communications" and "I Communications", respectively, General Chair of the 7th International Symposium on Wireless Personal Multimedia Communications (WPMC'04).

He spent the summer of 1985 at AT & T Bell Laboratories on leave from the University of Padua, doing research on Digital Radio Systems. He was Principal investigator for national research projects entitled "I bit rate mobile radio communication systems for multimedia applications" (1997–1998), "I Systems with Applications to WLAN Networks" (2000–2002), and "I-CDMA: an air interface for the 4th generation of wireless systems" (2002–2003). Also, he Task leader in the FIRB PRIMO Research Project "I platforms for broadband mobile communications" (2003–2006).

He is actively engaged in research on Digital Communication Systems and Brain Communication Interface for motor rehabilitation.

Giulia Cisotto was born in 1985. She received the Master degree in Telecommunication Engineering at the University of Padua in 2010. Since then, she has worked on the EEG signals processing for several clinical applications such as the diagnosis of Amyotrophic Lateral Sclerosis, the detection of awareness changes in Persistent Vegetative State and Minimal Consciousness patients, and Brain-Computer Interface (BCI) for motor rehabilitation of upper limbs after Stroke. She had work for three years at the I.R.C.C.S. San Camillo Hospital Fundation at Lido of Venice (Italy). She is completing her Ph.D. (March 2014) and she is going to be enrolled at Keio University (Yokohama, Japan) as associate researcher since April 2014. Her current interest is focused on EEG signals processing for rehabilitation purposes, especially EEG-based BCI applications for motor rehabilitation of upper limbs.

Francesco Piccione, born on 10th July 1961 in Marsala (Italy). Neurologist, Head neurophysiology and spinal cord injury unit, DIRECTOR OF Neurorehabilitation Department at IRCCS San Camillo Hospital, Alberoni-Venezia, expert in rehabilitation for neurologic diseases (Brain Injuries, Cerebral Stroke, Multiple Sclerosis, Spinal Cord Injuies, Extrapyramidal and Cerebellar Diseases, Peripheral Neuropathies).

Adjunct Professor of Neurology in School of Neurophysiology technologists and in School of Physical Medicine and Rehabilitation, at Padua University. Publication of more than 100 scientific contributions in Neurology, Neurophysiology and Rehabilitation fields.

5

Quality Improvement of Generic Services by Applying a Heuristic Approach

Oleg Asenov[1], Pavlina Koleva[2] and Vladimir Poulkov[2]

[1]St. Kiril and St. Metodius University of VelikoTarnovo, Bulgaria
[2]Technical University of Sofia, Bulgaria

5.1 Introduction

In this chapter the specifics and problems related to the management in Service Systems (SS) without blocking of the service elements are considered. Some unsolved problems of the management in this special class of service systems are presented and analyzed. A model for dynamic pro-active service management without blocking of service elements is developed through the definition of two generalized parameters, Saturation range and Attraction range of the Service Element (SE). Based on this model a mechanism for slot-based pro-active management of a set of SE in a generic SS is detailed. An optimization problem is defined for the location of the service elements with pre-defined characteristics as subject to a generalized Objective Function – service of the incoming asynchronous requests, without blocking and lowest possible weighted cost of service. The optimization task is analyzed and transformed into a task of finding p-medians in a weighted graph having an unknown parameter - **p** and a specifically defined weighted Objective Function, related to the two generalized parameters of the proposed model. The optimization task is solved via the introduction of a modified ADD/DROP heuristic algorithm. The advantage of the proposed algorithm is its low computational complexity, fast convergence and effectiveness of the heuristic procedure. A sample application for the development of a model for the provision of efficient emergency medical care in a big city is presented. We consider that such a modified ADD/DROP heuristic service model could be suitable in many applications in the aspect of CONASENSE where real time pro-active

Convergence of Communications, Navigation, Sensing and Services, 91 126.

dynamic management is necessary, a requirement, which is characteristic for servicing in critical infrastructures and emergency management. Examples are systems without blocking in the fields of, power distribution, water supply, traffic control and telecommunications.

5.2 Service Systems without Service Element Blocking

Queuing models are applied in the classical SS or so called queuing systems with the purpose of system behavior evaluation under a dynamically changing set of users' requests. One or more queues are defined for each SE, which are controlled regarding a specific discipline and priorities. The situation is different if such models are applicable for modeling information service distributed systems [1], in the so called public services sector, such as telecommunications, water supply, power supply, and gas delivery. In the public sector the users or subscribers are expecting to receive the service immediately without their requests waiting in a queue to be serviced and without being interested in other users' activity and current system loading [2], [3]. In this regard, the SS architecture should provide a possibility of current monitoring of the set of requests and load repartition (dynamic association) with respect to the set of active SEs. The purpose is servicing of every user request to be provided without delay. The need of repartition or dynamic control of the correspondence between the two sets, i.e. the SEs set and the service requests set, is a consequence of the absence of a structure for buffering of the incoming requests.

Every SE in the class of SSs considered here has the property of limited performance, capacity and potential, or in general: the ability to process a finite number of requests per unit time [4]. In queuing systems when the servicing capacity limit is reached, the requests are arranged in a set of one or more queues with different priority. Unlike systems with queuing service control models, for systems without blocking when the performance limit is achieved, any further request to that SE will be rejected. When reaching the servicing capacity limit value in systems without blocking every new request has two alternatives:

- to be submitted again in a later randomly generated point in time (a mechanism applied when collisions take place in communication systems) [5];
- to be pro-actively redirected (dynamically associated) for servicing to another SE, whose own service capacity limit value has not been reached yet.

A classification of the SSs including control models is presented in Figure 5.1.

In this chapter systems with Pro-active Control of Resources (PCR) which are a class of the SSs without blocking will be considered. The idea of pro-active control by means of an implicit resources reservation shall only be effective if an applicable mechanism for optimal or close to optimal request association in respect of Servicing Points (SPs) can be proposed. The applicability of such a mechanism depends on the computational complexity and the convergence of the respective dynamic association algorithm.

For a greater part of public SSs delaying the request for service to a later point in time is an unacceptable alternative for the users. Most of the real SSs such as the ones in the telecommunications sector and utilities sector do not presume the possibility of creating and controlling queuing service models. For example, a power supply or power distribution network, forms a SS in which a "queue" is technologically inapplicable as an architectural element, as far as on the one hand electrical power cannot be stored, and on the other if the power requested is higher than the provided one, an "overloading" condition is reached. Quite similar is the example of a real time telecommunication service request. If the request is "queued" to be serviced later, the user can lose interest in the service to be provided, as it is needed only with a "real time" attribute, i.e. immediately.

Application of so-called hybrid control models is possible for the water sector, gas distribution and fuels supply [3]. With these models queuing structures with particular disciplines and servicing algorithms are defined for

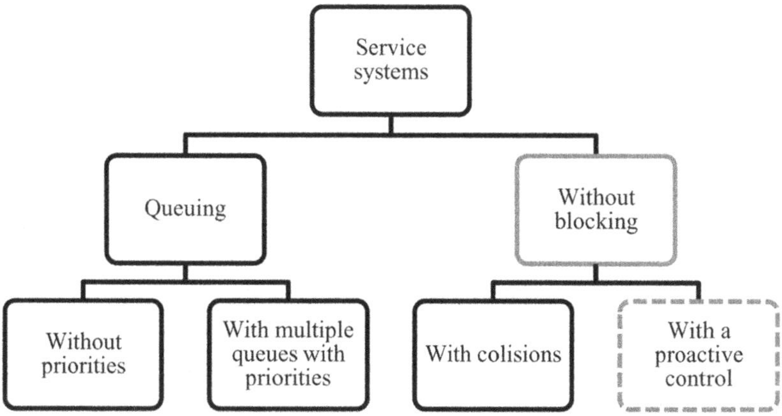

Figure 5.1 Classification of Service Systems Including Control Models

a part of SPs or SEs, while for others a "direct" servicing logic is applied, i.e. with non-blocking SEs. Regardless of whether the SS permits control with blocking of a SE or hybrid control of the SE, the problem of modeling and studying of systems in which the SP are modeled without a queuing structure and the SE is of the non-blocking type is up to date due to the following considerations:

- obvious practical applicability;
- arising need of control mechanisms with a real time applicability and low computational complexity;
- flexible control reflecting user requests dynamics;
- system management aiming at minimization of the servicing costs while ensuring a requested Quality of Service (QoS).

The proposed solution of the problem defined is by means of the application of *methods for pro-active control with dynamic association of SEs to the set of user requests for service*. The application of methods with dynamic association of SEs to the set of user requests for service is possible only under the condition that the system architecture allows request redirection from one SE to another without affecting substantially system reaction time and performance or QoS.

The basic idea of the methods which include dynamic association of the SEs is based on the assumption that their system architecture can be adapted by means of a specific control action which we call Dynamic Control Impact (DCI). From a practical point of view the problem can be reduced to the following statement: through proactive system control we want to achieve a Service Oriented Architecture (SOA).

The goal of implementing a DCI is to provide service without buffering the user requests for service within a finite time interval, called by analogy to telecommunication systems "a slot" (Figure 5.2).

Application of such slot-based control mechanism in SSs is an established practice in the field of telecommunications such as slotted IEEE 802.15.4n, but the standards are developed either on basis of queuing models (Figure 5.1) or on control models without blocking of the SEs with discovering and solving collisions (for example slotted ALOHA), for which new applications are constantly being developed [6]. In this work the slot mechanism is applied with the purpose of the proposal of a SS with PCR without blocking of SEs.

For SSs with PCR the slot is a time interval within which the "service request – servicing element" (user – service point) ratio is established by means of an association table (Figure 5.2). The association table is a relation of the distributed database for supporting the dynamic association mechanism

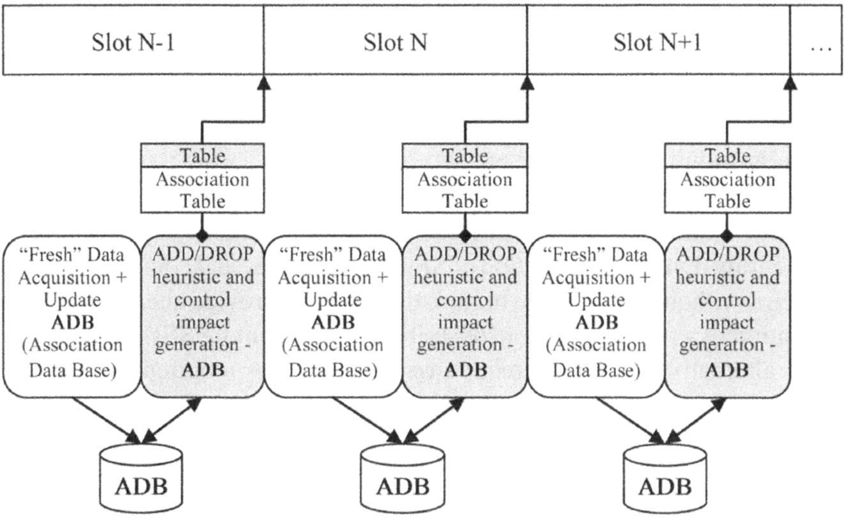

Figure 5.2 Mechanism for Servicing Without Blocking and Pro-Active Resource Control

or so called AssociationDataBase (ADB). From practical viewpoint the DCI is concretized in the association table. The latter is one more architectural element, which to a great extent has an analogue in telecommunications, the so-called ALOHA Reference Table [6]. The meaning of the ALOHA Reference Table is to establish the access rights by means of a communication slots distribution map. But in contrast with ALOHA where not every request should necessarily fall within the distribution map, in the class of SSs without blocking and PCR, the request association rejection takes place in absence of service.

The development and studying of the application properties of the methods for dynamic association for a class of SSs with PCR is related to the definition and analysis of:

- a characterization model of the servicing properties of the non-blocking SEs;
- a dynamic association algorithm which handles the process of dynamic definition of the number, location and properties of the active SEs in relation to the asynchronously incoming service requests;
- a dynamic association algorithm for the asynchronously incoming service requests to the current active set of SEs with specific properties;
- an information model of dynamic association methods for a class of SSs with PCR which includes:

- a data model for support of the PCR algorithm and using a distributed relational database with centralized management called ADB (Figure 5.2);
- a model of the applied system processes for support of the PCR algorithm.

The architecture of telecommunications, power distribution, power supply, water supply or traffic control networks to a great extent contain the possibility of implementing dynamic association of SEs. The topology and connections in these systems without blocking are built so that they can provide alternative for the association of a request (user or subscriber) to two or more SEs. Provision of service alternativeness is a direct consequence of the requirements, laid down at the designing stage, for guaranteeing a minimum Out of service (OoS) time. This is a requirement, which is characteristic for most SS and in many cases is part of the so called critical infrastructure [7].

From a theoretical viewpoint all developments in the field of the Queuing Systems are applicable to the class of SSs with PCR under the assumption that the queue is with a dimension of "one". The problem from an application aspect is that the interaction between the two probabilistic processes, the generation of service requests and servicing of requests, is based on probabilistic models such as Markov networks and Poisson processes. Investigations lead to models where the control itself is related to analysis and evaluation of the service delays depending on the intensity of users' requests and the choice of an effective configuration and discipline for the control of one or more queues. This is a reactive approach as it evaluates system behavior under changes in loading. The computational complexity of the procedures is so high that it does not allow "real time" application. For long term planning of SSs resources this approach is efficient and has its advantages.

In the case of dynamic association the model actually "reverses". The ratio request and servicing element, as well as the number and outputs of the SEs, are uniformly distributed within a limited time interval, that is a slot. Within this slot the control decision is made, practically appearing as a new system service profile, which is valid for the next slot (i.e. the new request to servicing element ratio, number and outputs of the SEs) as presented in Figure 5.2. The so defined model sets an additional requirement to the service provision mechanism of the system, in which the dynamic association method is applied. This requirement could be defined as a semantic and contextual independency of the requests for service from the respective SE. In this sense if the dynamic association logic defines such a type of managing impact, then when an application request has been started for servicing to or by a SE within the current slot, the "transfer" or redirection of the servicing of this request to another SE during the next slot must be possible. This requirement is important as an applicability criterion of the methods reviewed, but it is not limiting for the typical classes of application systems for which these methods are applied. In general we know, this is a key functionality that characterizes these systems; in telecommunications this functionality is the so called roaming, and in power transmission networks it is called an alternative switching contour.

Expectations of the so-defined methods of dynamic SEs association to the set of the users' service requests suppose a formation of a control impact within each slot. This should define a new version of service system structure so that the following goal is met:

"service of user requests with equal or better than the requested QoS while minimizing servicing costs".

In view of the methods of dynamic SEs association to the set of users' service requests we can define the so-called "ideal" control impact. This is the one that within each slot provides such an architectural profile of the system, in which each request is placed in compliance (associated for servicing) to a SE with an equal or larger free capacity than the intensity of the requests for services. Closest to this "ideal" architectural profile is the service model of the "1:1" type, i.e. the number of SEs is equal or greater than the maximum number of active user requests and for every request at least one SE can be associated.

This, from a hypothetical point of view, seems analogue with the case in telecommunications where every mobile user (subscriber) in a cellular network is provided with exactly one base station which serves him only. The resources of this station are such that they provide the maximum service performance ever requested by this subscriber. If we assume in the example revealed that the subscribers are fixed it is possible to place a base station in the vicinity of each subscriber and thus always every servicing request will be serviced, but the servicing price will be the highest possible. In this aspect cellular networks with mobile subscribers could be considered as one of the best examples of a system in implementing a servicing logic for the realization of the 1:1 dynamic association principle.

For real systems in which dynamic association can be applied as a service control model it is typical that the SEs subsystem is determined as a number, disposition and SEs properties while the users' system is dynamic regarding the parameters characteristic for the requests servicing process. In this case from a SEs viewpoint the dynamics of the architecture is expressed through the definition of activity profiles for every SE. The activity profile determines the servicing properties of the element. If for the next slot no service requests that could be associated to a specific SE (service point) are foreseen, then it may turn into "passive" (or stand by) profile. Thus as a result of the application of the PCR methods an additional "green" effect for the whole system could be obtained, i.e. energy expenses reduced for SEs with limited servicing properties and/or SEs in a "passive" mode. The results of the analysis of different versions of the system architecture for a continuous sequence of slots may also be used for the long term planning of the topology and functionality of the SE.

Within a current slot (Figure 5.2), it is expected that the dynamic association logic will produce a control impact which appears as an instruction (Activity Table), showing the possible activity profiles that a specific SE could have for the next slot. The Activity Table is a function of the Association Table. Based on the expected volume of requests for service calculated by means of the Association Table for each SE an efficient servicing mode can be chosen.

5.3 Dynamic Association as a Management Approach

As management approach, the dynamic association considers the Service System's life cycle as a continuous sequence of slots. Within the frame of each slot a control impact is applied, which determines the Activity profile of each SE and the compliance (association) of the SEs sets with the current active set of requests for service.

The slot principle of predicting and decision making about the character of the control impact is of key importance for the dynamic association as a management approach. The input information for determining a new state of the system is:

- the statistical feedback about users' activity during the current slot;
- the assumption that this activity will not change substantially within the next slot;
- the activity profile of each of the SEs for the current slot;

The control impact is a result of an optimization procedure for satisfying an Objective Function of the dynamic association, i.e.:

"servicing user requests with equal or better than the requested QoS while minimizing servicing costs".

Slot duration is an important parameter of the dynamic association approach (Figure 5.2). On the one hand slot duration has to be commensurate with the time interval within which the profile of the user requests is relatively constant, while on the other hand the slot has to be long enough, so that from practical viewpoint it provides a sustainable enough association of a user to a specific SE. In this sense the slot duration depends on the application characteristics of the system. In mobile communication systems, for example, slot dimension is expected to be substantially smaller than the slot in gas distribution or water supply systems.

A possible approach for a general solution of the dynamic association problem for SS with PCR is:

- to develop and study a general model for presenting the servicing process;
- to define a generalized Objective Function.

It is important to note that for the model definition we should consider the fact that in order to be applicable this model should reflect also the specifics of SS with PCR as given in next:

operability; meaning relatively little time and limited resources for the analysis and definition of a control impact;

applicability; meaning a control impact to be interpretable, i.e. it has to be implemented in the existing system architecture without substantial changes in the individual management of the SEs;

scalability; the model should be applicable regardless of the number of users and SEs;

comprehensiveness; the model should reflect the parameters that are essential for the service and also those for the requests, as well as the provision of their implementation;

adaptability; the model should be general enough so that it is applicable to various types of dynamic service systems;

feasibility; the model should be oriented towards an objective control function, clarifying the type and character of the control impact applied to the SEs and/or user means of access to the services;

"fair play"; the model should provide a possibility for evaluation of the management efficiency in relation to the SS operators as well as to the users requesting a service, without breaking the agreements on servicing levels and minimum average OoS time.

The scheme of the dynamic association (Association Table) and control of the SEs network (Activity Profile Table) as well as the respective procedures and rules for the practical association and control of the SEs network can be determined by optimizing the generalized Objective Function.

5.4 Characteristic Model of the Servicing Properties of the Non-blocking Servicing Elements

The model is based on the empirical idea about the market positioning of a service by means of the service provision price from user's point of view as well as on basis of the costs for provision of the same service from the viewpoint of the provider (operator ensuring the normal operation of the SE).

For every user the need for a certain service presumes finding clear answers to the three questions: "Where, in how much time and at what price shall I receive this service?". The integral minimum of these three aspects would determine the natural choice of a SE. Depending on the character of the arising necessity of servicing the user may prioritize or neglect some of the answers to the questions. In other words, when minimizing the objective function it is acceptable that weight coefficients are introduced which should determine the character of the arising necessity of service in case of requests from different users for one and the same service. This means that when choosing a SE, the type of service requested as well as the additional conditions characterizing the user's profile should be evaluated.

Planning the positions (topology) of the SE is a basic problem for a service operator. When solving this problem a number of additional, non-technological constraints and special features are usually introduced, which are not subject of studying and modeling, i.e. it is assumed that possible and impossible (inefficient) topologies exist and such logic is implemented when choosing the topology. Operators practically use an approach which is characteristic for the user. In order to position a system of SE with a specific capacity the operator usually bases his decision on a statistical evaluation of the arising necessity for provision of one or a group of connected services. In the evaluation geographical bounds can be represented by zones with different properties with respect to intensity and profiles of service requests. If is the requests are based on statistics (as for consumption of foods or basic products) the statistical zoning approach works fine, but when the system's dynamics is high and the zones currently change their properties, the limits of the zones are quite conditional and they practically change dynamically. In this case the statistical data must be "refreshed" in order to reflect the current condition and zoning should be considered as function of time with validity equal or commensurate with the duration of the service slot (Figure 5.2) characteristic for the system.

When composing the model it is important to estimate the acceptable constraints for the time of control decision making. If we assume that a system is reviewed in three consecutive time slots (Figure 5.2) and the slot size is reduced so much that it provides relatively constant properties of user service requests, a successful management approach presumes statistics to be accumulated during the time of the first slot, based on which an estimation of the service profile for the next slot is made and a control impact is chosen. For such a solution it is assumed that there is correlation between the service requests profiles in two adjacent slots i.e. the difference is insignificant and the preceding slot outlines the tendency which develops at least in the actual slot. However, the shorter, the time for data processing and choice of a control decision, the longer statistical data will be accumulated and more exact will be the probability estimation for the user requests' intensity for the next slot. In this sense the development of low computational complexity algorithms, fast convergence and minimum configuration recalculation time is an important task for a successful application of the model reviewed.

When defining the model it is assumed that for every SE we can define at least two basic properties, which are represented it in the process of pro-active control with dynamic association:

Attraction rate. The property is needed in order to answer the question: at what distance from a user can this SE provide service? Here by distance we mean an integral evaluation of the remoteness - range, time to reach, price to reach, etc. Different values can be attributed to this property for the different SEs as a meaning of the property itself. For example, if the SE is a base station of a mobile network, it has a coverage zone within which a mean value can be defined for the most distant point from which a user (in this case mobile subscriber) can receive service from this base station. If the SE is an Ethernet network switch and users are computers connected to it by a physical environment twisted pair 100BaseTx, then the value of this property could be 100 meters. If a user is connected through a cable over 100 meters long from a network switch, then this user cannot be served by this switch. If the SE is a base station which services cable modems in a cable operator network - then the maximum distance from which a cable modem can be apart from the base station is determined depending on the cable type and amplifier installed. It is obvious that the *attraction rate is a property related to the ability of service provision.*

Saturation Rate. This property specifies the maximum amount of user requests with certain properties which a SE can service without waiting on a queue for servicing. The saturation rate is connected to the SE performance (output). It is important to note that the considerations are made for SS in which there is no queue for storing service requests received (without blocking), or a possibility for some type of priority control. Saturation rate is thus related to the *capacity of servicing of simultaneous requests*, possessed by the SE.

Based on these two properties of the SE it is possible on the one hand to present and evaluate the efficiency of every dynamic SS given the topology of users and the structure of user requests. On the other hand, on basis of such a model of presenting the SEs the straightforward problem can be solved as well. Given (based on empirical and statistical evaluation) a users' set as topology and as structure of requests, an efficient SS structure should be defined. Under these constraints the servicing architecture can be controlled dynamically by configuring a version of the structure with a respective number of SEs within each time slot. Thus, a possibility is provided to satisfy the condition for an adequate system control, i.e. providing user service with the required QoS while minimizing servicing costs.

In practice the problem solved is the "intermediate" problem. *Given the topology and properties of SEs, and data on the evaluation of the users set as disposition and character of the servicing requests, we are able to find an approach for a dynamic management of the current architecture and a way to estimate the development of the architecture in the long term.*

Formation of the Objective Function for solving the "intermediate" problem is based on the constraints in relation to the model and the heuristic prognosis that the measures determined in the Objective Function, represent the process objectively and service the objectives of the participants in this process, in the case of dynamic SSs without blocking. We assume hereby the constraint that the model will be developed by the apparatus of solving linear programming problems and with respect to minimizing the computational resource costs.

The starting point for developing the model is based on the solution of a classical graph theory problem, i.e. the problem of finding P-medians in a weighted graph [8]. From the viewpoint of the constraints applied on the model a new algorithm has to be developed and/or an existing one has to be modified combining the advantages of the Heuristic Concentration, Linear Programming and potentials in Graphs Theory to executing relevant tasks [9]. Firstly, Heuristic Concentration is an applicable approach for developing the servicing mechanism that is typical for SSs without blocking. Secondly, the constraint of low computational complexity presumes linear programming as an approach for defining the association Objective Function. Thirdly, the graph theory apparatus is obligatory from a systematic point of view as long as the graph theory models are characteristic for the object discussed.

The method proposed is based on the two basic properties of the SEs already defined – *attraction rate and saturation rate*. In this model the evaluation concerning the attraction rate is represented from the graph theory viewpoint by the *weight matrix of connectivity*. The weight of each arc is calculated in the same metrics in which the property attraction rate is measured. Thus based on the weight matrix of connectivity an evaluation of the service zone borders of each point is performed.

The saturation rate property is represented by the *weight matrix of the nodes in the graph* which represents the SS. The weight of each node is a positive number corresponding to the saturation rate of the SE (in the respective metrics for capacity, output or another applicable quantitative measure of the limit of element blocking), or zero, if the node is a user with service requests. In this way two types of nodes are defined in the model – serviced and servicing ones represented respectively by zero or non-zero value of the property saturation rate.

It is important to note here that in the case of "0" requests for the next slot the system has a constant own residual value and this is the price paid for the existence of all SEs (servicing nodes in the graph). This *static price component* is time independent and is a function of the system scale, i.e. the number of SEs installed.

This *dynamic price component* for every specific slot (Figure 5.2) of the system's life cycle depends on the:

- current active set of subscribers serviced;
- current version of the servicing architecture represented by the Association Table (Figure 5.2) and the recalculated one based on the Service Points Activity Table.

For the formation of the connectivity weight matrix an evaluation function has to be defined, which should reflect the specifics of the functioning mechanism of the type of SSs. The weight matrix of the graph nodes gives an evaluation of the relative price of the infrastructure, which provides service to users within the current time slot. The evaluation depends on the value of the Activity Table element corresponding to a service node in the graph. This matrix reflects simultaneously the static and the dynamic components of the price for service, which within the slot gives an estimation of the load dependent and/or independent servicing expenses generated by each point.

The variable in the model and respectively its formal presentation as a linear programming task is the Association Table or the Association Matrix. The Association Matrix is a two dimensional matrix with a ***NxM*** dimension where N is the number of all the nodes in the graph and M is the number of servicing (active) nodes in the graph. The association matrix provides unequivocal and current information regarding: "from which SE the requests of each user within the next slot will be serviced". From practical viewpoint the Association Matrix is a physical expression of the dynamic control impact and a current version of this matrix is generated for every slot.

Key element of the model is the definition of the *Objective Function which gives the price for service which subscriber i pays to be provided with the requested services from SE j remoted at a distance* d_{ij}*from subscriber i.* The following parameters are introduced during the formation of this Objective Function:

- Fixed price of the SE. This becomes a static component if it does not service users;
- Operational price of the SE. This becomes a dynamic component function of the aggregate services requested by user i, serviced by point j;
- Servicing price of the SE. This becomes a dynamic constituent giving the price which user i pays for the currently requested services if he is serviced by point j. This is the price for the infrastructure and other direct costs in the process of providing access of the user to the SE he is associated to. The price is determined mainly on the basis of the remoteness of the user from the SE.

For modeling purposes we assume that this price is infinitely high when a user is situated at a distance bigger than the saturation rate limit of a specific SE, despite the fact that the infrastructure can provide formal association to that SE. The Objective Function is defined as:

$$\begin{aligned}ObjectiveFunction \;&= PriceforServicing(i,j) + Pricefor\\ &Operation(i,j) + PriceoftheSE(j)\end{aligned} \tag{5.1}$$

This function reflects the heuristic assumption for the users' behavior when choosing a SE which is determined by the criteria: "fastest and cheapest" for one and the same or commensurate quality of the service provided. The two dynamic components, servicing price and operation price, are a function of the control impact (Association Matrix). The static one, price of the SE, is assumed to be one and the same within each slot and depends only on the installed SE number. The first term of the Objective Function reflects the servicing system dynamics in view of the costs that users pay to receive the services requested under a servicing scheme determined by the Association Matrix, describing which user is serviced by which SE. The second and the third terms represent the operation price of the SE and its adjacent infrastructure depending on the user requests serviced, both are numbers and are also a result of the association.

The goal of solving the optimization problem is to find the type of the association matrix (variable matrix in the problem), where the Objective Function value reaches its minimum and where conditions (2.2) and (2.3), which are a consequence of the practical constraints in the functioning of a SS with PCR, are satisfied,

$$\sum_{i=1}^{N}\sum_{q=1}^{K} S_{iq} \leq P_{s_j}, \quad for\ \forall j \in 1 \dots M, \tag{5.2}$$

where $\boldsymbol{Ps}_j$ is the saturation rate for SE j and

$$\sum_{i=1}^{N}\sum_{j=1}^{M} a_{ij}d_{ij} \leq P_{a_j}, \quad for\ \forall j \in 1 \dots M, \tag{5.3}$$

where $\boldsymbol{Pa}_j$ is the attraction rate for SE j.

The solution of the problem is found by deriving such a type of Association Matrix $[a_{ij}]$ of the actual control impact (map of the dynamic association of users to SE for the next slot), where the requested QoS is achieved at the lowest total servicing price.

The application of such a proactive management scheme for a system of distributed SEs providing public services has the potential to ensure:

- a possibility for reduction of the total OoS time for the system;
- an increase of the operational efficiency by managing the number of active SE for a measurable time interval (month, week, day, hour, second etc.);
- effective planning of the system development in the short and long term – adding new points, changing properties of existing points ($\boldsymbol{Ps}_j$ and $\boldsymbol{Pa}_j$);
- application of energy efficient ("green") control policies of the system of SE;
- possibility of performing simulations and analyses of different situations of failure in a SE and access infrastructure and the generation of emergency versions of the Association Matrix.

The development of this model and solving such class of problems has a measurable application value when and only when the service provision architecture ensures alternativeness of service, i.e. at least two or more SEs exist for one user request within the attraction ratio zone at the moment of request sending. When, after applying the saturation rate criterion for a user, only one SE is accessible for service provision, then for this system the solution of the so defined class of problems is inapplicable in its basic sense, i.e. service optimization. In modern competitive models of service provision an alternativeness of the SE choice is available. The topicality of this class of problems is outlined even to a greater extent by conducting policies of infrastructure consolidation and virtualization and provision of the possibility of more than one operator to install and support alternative SEs systems upon a single physical infrastructure.

Depending on the system dynamics, service requests distribution and intensity an additional group of constraints or limitations is formed. Limitations concern:

- ΔT, slot duration within which a single value of the Association Matrix variable $[a_{ij}]$ is applied;
- ΔT_{cal}, time for calculating the new value of the Association Matrix variable $[a_{ij}]$;
- Δ_{man}, time for distribution and application of the new value of $[a\Delta_{ij}]$.

Based on these constraints and the natural requirement for continuity of servicing, the time interval for the statistical evaluation of service requests distribution and intensity ΔT_{stats} is determined from:

$$\Delta T_{stats} = \Delta T_s - (\Delta T_{cal} + \Delta T_{man}) \tag{5.4}$$

The following conclusions can be drawn from expression (2.4). A greater time interval ΔT_{stats} is required for obtaining the prognosis on the service requests distribution and intensity to be reliable and applicable. We note that ΔT_s is a constant for every SS and is dependent on the technological constraints related to the possibility of application of policies of alternative users association to the currently accessible SEs. The same may be true for the time interval ΔT_{man}, which is of a technological nature and is thus independent from the association process. In order to maximize ΔT_{stats} under these conditions, only one possibility exists to minimize the time interval for calculation of the new value of $[a_{ij}]$ and this is to reduce ΔT_{cal}. In this respect problem solving focuses on the development and application of an algorithm with a relatively low computational complexity, guaranteed convergence and efficiency for minimization of the objective function. The solution determines the applicability of the model presented.

From Graph theory viewpoint seeking of the minimum of an objective function of the type:

$$Z = \sum_{i=1}^{N} \sum_{j}^{P} a_{ij} w_{ij}, \tag{5.5}$$

is related to the solving of the so called problem of finding p-medians in a weighted graph [8]. Here $[a_{ij}]$ is a matrix of association of nodes i to j; a_{ij}= 1 presents the fact that node i associates to node j (a_{ij} = 0, node i does not associate to node j); $[w_{ij}]$ is the weight matrix of the undirected graph $G(N)$.

The problem of finding p-medians in a graph is central to many problem classes, which can be found in the literature under the name "distribution and disposition of servicing centers" [8–14]. When introducing the estimation of the weight for each vertex of the graph (not only for the arch as in equation (2.5)) the classical p-median problem turns into the so-called generalized problem. When solving the generalized problem, a fixed price F_i is attributed to each median vertex [15].

The generalized problem of finding a p-median can be formulated as follows. For the given graph $G=(X, A)$, with shortest distances matrix w_{ij}, with vertex weights V_i and with fixed vertex prices F_i, the problem consists of finding such a subset $\overline{X_p}$ of p vertices, for which the equation:

$$Z_G = \sum_{i=1}^{N} \sum_{j \in \overline{X_p}}^{P} a_{ij} w_{ij} + \sum_{j \in \overline{X_p}}^{P} a_{jj} F_j \tag{5.6}$$

has the minimum possible value.

When modeling real SSs with PCR the introduction of a fixed price of the SE does not reflect the system's dynamics and the difference in the resources expenses, which arise from the servicing of a various number of users and user requests by one SE. With the purpose of a more exact evaluation and better formulation of the Objective Function, it is assumed that the price $F_j(\Delta t_s)$ of the i-th SE within the servicing slot (Δt_s) consists of two parts:

$$Z_j\left(\Delta t_s\right) = F_j + Fd_j\left(\sum_{i}^{N} a\left(\Delta t_s\right)_{ij} w_{ij} \overline{S\left(\Delta t_s\right)_i}\right) \tag{5.7}$$

In expression (2.7) $S(\Delta t_s)_i$ are the requested services from SE i during the time slot Δt_s.For the purpose of modeling the class of SS with dynamic association of the SEs to the set of users' service requests a generalized Objective Function is defined by substituting the expression for $F_j(\Delta t_s)$ from expression (2.7) into expression (2.6). This objective function is now given as:

$$Z_G(\Delta t_s) = \sum_{i=1}^{N} \sum_{j \in \overline{X_p}(\Delta t_s)}^{P(\Delta t_s)} a(\Delta t_s)_{ij} w_{ij}$$

$$+ \sum_{j \in \overline{X_p}(\Delta t_s)}^{P(\Delta t_s)} a(\Delta t_s)_{jj} \left[F_j + Fd_j \left(\sum_i^N a(\Delta t_s)_{ij} w_{ij} \overline{S(\Delta t_s)_i} \right) \right] \quad (5.8)$$

In many cases in which the problem of finding the p-medians is solved, the value of p is known in advance, i.e. before performing the association and displacement of the p-number of SE [8].

When minimizing the Objective Function in expression (2.8) the number of SEs *is not determined in advance* and it is being determined for each slot depending on the concentration and intensity of users' service requests. This model is efficient from an application point of view as far as many problems in practice suggest the possibility of dynamic "on and off" switching of SEs. Such examples are mobile networks where one of the practical mechanisms for reducing costs is the switching off of base stations in the service range of which there are no users. Similar mechanisms for an efficient control of the SEs are applicable in the utility operators as well as in building automation systems. On the other hand, when designing and/or planning a SS, one of the most frequently asked questions is: "how many, what type and where should the SEs be placed?"

By simultaneous application of the developed model and the minimization of the Objective Function in expression (2.8) are solved the key issues related to the configuration dynamics of the SS with PCR:

- determining the number and place of the active SEs within the next slot;
- association table (map) of user requests for the next slot, i.e. which user will be serviced by which SE.

5.5 Heuristic Algorithm for Dynamic Association of Asynchronous Requests

For the purpose of multi-parametric optimization of the Objective Function in expression (2.8) it is necessary to develop an association algorithm in order to ensure a possibility for simultaneous calculation of the value of p

and determination of the association matrix $[a_{ij}]$ at which $Z_G(\Delta t_s)$ reaches a minimum value. Practically this will be an algorithm for dynamic definition of the number, place and properties of the active SEs and a mechanism for slot-based association of the asynchronously entering user requests to the current active set of SEs with specific properties. After review of the sources from [8] and [9] the conclusion may be drawn that when defining the problem, the value of p is known in advance and the subset of nodes in the graph with a dimension p is calculated, for which the Objective Function (defined for the purposes of the specific application) reaches a maximum or a minimum value.

Here we propose a heuristic algorithm for a double parametric optimization of the Objective Function (2.8) which within the sense of the applied heuristic procedure is called ADD/DROP heuristics. ADD/DROP heuristics aims at defining such an absolute medians set in graph $G=(X,A)$with dimension p and terms$\overline{X_p} \in X$, where the generalized objective function from the expression (2.6) reaches its absolute minimum value [10].

From practical viewpoint for the SS with PCR without blocking when developing the ADD/DROP heuristics algorithm the following is taken into consideration. When composing the model it is possible to reach a situation that the objective function should be defined for the case of solving a problem for finding a generalized p-median [11], [12]. This problem unlike the typical p-median problem considers the possibility that the number of the median vertices $[\overline{X_p}]$ can be less than p [13], [14]. The problem for the minimization of expression (2.8) when $[\overline{X_p}] < p$ is a variation of the problem for finding a p-median and it is quite common in practice [15].

5.5.1 Formulation of the Problem for Finding a Generalized P-median Set as a Linear Programming

Consider a weighted undirected graph $G = (X, [a_{ij}])$ with a set of vertexes X and set dimension of this set $[X] = N$ and association matrix $[a_{ij}]$, which represents the distribution of the set of non-median $\overline{X_n}$ to the median vertices$\overline{X_p}$, while fulfilling the condition:$\overline{X_n} \in X \bigcup \overline{X_p} \in X;\ \overline{X_n} \bigcup \overline{X_p} = X.$

When composing $[a_{ij}]$ the following rule applies:

$$a_{ij} = \begin{cases} 1, \ if\ vertex\ i\ is\ associated\ to\ median\ vertex\ j \\ 0, \ if\ vertex\ i\ is\ not\ associated\ to\ median\ vertex\ j \end{cases} \tag{5.9}$$

When defining the Objective Function it is assumed that:

$$a_{ii} = \begin{cases} 1, \ if\ vertex\ i\ \in\ \overline{X_p},\ i.e.\ i\ is\ a\ median\ vertex \\ 0, \ if\ vertex\ i\ \in\ \overline{X_n},\ i.e.\ i\ is\ not\ a\ median\ vertex \end{cases} \tag{5.10}$$

The variables when solving the problem are:

- *P*; dimension of the median set;
- $[a_{ij}]$; type of matrix;
- $[X] = N$; dimension of the median set.

The constraints are:

$$\sum_{j}^{N} a_{ij} = 1, \tag{5.11}$$

$$\sum_{i}^{N} a_{ij} \leq P_{s_j} \in \overline{X_p}, \tag{5.12}$$

$$a_{ij} w_{ij} \leq P_{a_j}, \quad i \in \overline{X_n}, \quad j \in \overline{X_p}. \tag{5.13}$$

These constraints result from the model presented in section 1.4 and they reflect the performance characteristics of the class of SS with PCR:

- Each user /request is serviced by one SE (2.11);
- The SEs in the system are without blocking and associate a number of users/requests not exceeding the individual saturation rate Ps_j(2.12);
- Each SE provides servicing of users/requests at a relative weighted distance not exceeding the individual attraction rate of point Pa_j (2.13);

The formulation of the problem in this case is related to finding the minimum of the Objective Function

$$Z(p, A) = \sum_{i=1}^{N} \sum_{j=1}^{N} a_{ij} w_{ij} + \sum_{j=1}^{N} \left(f_{d_j} + f_{s_j} \right). \tag{5.14}$$

A special feature in the composition of this Objective Function is the second term, which reflects the median center weight. The specific evaluation of the weight through the introduction of a dynamic component fd_j is a characteristic evaluation for this class of systems. The statement is that the price of the SE within the current slot is a function of the users (requests) association. The validity of this statement is supported by empirical facts, showing that the concentration of requests for service suggests a more "expensive" SE without blocking. The addition of a dynamic component in the weight evaluation of the SE is of key importance for minimization of the objective function (2.14). This is justified from the following assumption. If we accept even only theoretically that the system can function with only one

SE, then on the one hand this point will have maximum capacity and respective price. On the other hand, the price of the infrastructure will also be maximum. In the other limiting case – when every node is also a SE the price of the SE will be minimum because only one source of requests will be serviced. But the infrastructure on the next hierarchical level will have an additional price. This is due to the fact that with cascade systems if we decide to lower the costs excessively at one level this immediately reflects on the costs of the next higher level. In this sense, solving the "planar" problem for finding and disposing p SEs points with values of p and $[a_{ij}]$ for which the objective function of (2.14) reaches a minimum does not solve entirely the problem of choosing the architecture for the SS with PCR. This requires the additional formulation of a cascade problem for finding connected sets of p-medians.

5.5.2 ADD/DROP Heuristics Algorithm

Development of heuristics procedures with gradual adding and reducing of the number of median vertices is a practice in research related to quality improvement of tracing ray algorithms [15]. In this chapter the idea of reaching a close-to-optimal median set by gradually adding/reducing the number of median vertices develops into two reversive algorithmic sequential procedures: ADD and DROP.

In the ADD procedure assumption is made, that the disposition of the SEs (p-medians) starts with one single SE placed in a median of a specific weight graph, i.e. in the node which is closest to all other nodes. If the so-called Floyd matrix [15] is calculated for the weight graph, the median is the node (vertex) for which the sum of the weights of the Floyd matrix rows is smallest, i.e. this is the node at the smallest weight distance from all remaining ones. Based on the heuristic assumption this node is accepted to be the first most probable member of the median set. The ADD procedure operates with a continuously increasing median set until a dimension p is reached at which the objective function from expression (2.14) is of the lowest value. At each step a new median node is added, assuming that a solution close to the optimal one is reached for such a value for **p** for which the following inequality (2.15) is valid:

$$Z_G(\Delta t_s, (p-1)) \geq Z_G(\Delta t_s, p) \leq Z_G(\Delta t_s, (p+1)) \tag{5.15}$$

The sequence of steps in the ADD-procedure of the ADD/DROP algorithm is as follows:

Step 1: Choice of a starting node: p=1. The starting node is the median node in graph G=(X, $[a_{ij}]$), with a matrix of the shortest distances w_{ij}(Floyd Matrix). This is the node with an index i, for which the following condition is valid:

$$i \in \overline{X_p}, \quad \sum_{j=1}^{N} w_{ij} \to \min \tag{5.16}$$

Step 2: Choice of a new median node and successive addition of each of the remaining [N–1] nodes as a potential member of the median set $[a_{ij}](\Delta t_s)_{ij}$.

At this step N-1 versions of the Objective Function with a dimension of the median set p=2, $Z_G(\Delta t_s$,p(2) are calculated. A corresponding type of the association matrix is composed for each version of the median set. The principle in composing of the association matrix is the "distance" of the node to another node member of the median set. The association rule is applied until the composition of a full version of the association matrix $A(\Delta t_s)_{ij}$ is reached.

Having composed an association matrix for $a(\Delta t_s,p(2)_{ij}$ the intermediate value of the Objective Function is calculated for the current set of the median nodes according to (2.5) and (2.14), i.e. $ZG_{p(2)}(\Delta t_s,p(2)$.

Amongst the N - 1 candidates for a potential second member of the median set, that second title member of the median set for the i-th node, is accepted the one for which the following condition is satisfied:

$$p\,(2) \in \overline{X_p}, \quad \sum_{j=1}^{N} Z_{G_{p(2)}}\,(\Delta t_s, p\,(2)) \to \min \tag{5.17}$$

Step 3: The median set increases by adding new terms in accordance with Step 2 until the condition in (2.15) is fulfilled, which is actually the condition for an ADD procedure exit.

As a result the current value of p (number of SE) is determined. At this value the objective function $ZG_p(\Delta t_s,p)$ has its minimum.

The DROP procedure is a reversive procedure to the ADD. This procedure starts with a median set with N terms, the number of terms being reduced with one at each step until reaching that value **p**, for which a solution which is closest to the optimal one is reached under the condition of

$$Z_G\,(\Delta t_s, (p+1)) \geq Z_G\,(\Delta t_s, p) \leq Z_G\,(\Delta t_s, (p-1))\,. \tag{5.18}$$

The specific functionality of the ADD/DROP heuristic algorithm consists in its operation with a median set with a dynamically increasing/decreasing number of terms. The association mechanism (according to the rule from Figure 5.2) is applied in both reversible branches of the algorithm. In case of a randomly generated weight graph with 50 nodes the application of the ADD and DROP procedure leads to the fulfillment of inequalities (2.15) and (2.18) for one and the same dimension **p** of the median set and one and the same result in respect of the value of the objective function. For the purpose of visualization of the algorithm behavior in view of its multi-parametric optimization properties an example of normalized results for an Objective Function is shown in Figure 5.3. The calculation is performed for a simplified version of the Objective Function as follows:

$$Z\left(p, A\right) = \sum_{i=1}^{N}\sum_{j=1}^{N} a_{ij}w_{ij} + \sum_{j=1}^{N} f_{s_j} \tag{5.19}$$

Here fs_j is a randomly fixed price of a SE, generated within the variation range of w_{ij} values. By doing so the effect of reassessment of one of the terms participating in the formation of the Objective Function is isolated.

The results from Figure 5.3 represent the functioning of the algorithm. At small values of **p** the value of the Objective Function is high, which is a result of the bigger weight of the access infrastructure to the relatively small number of SEs. The weight of the newly added points increases with the increase of **p**, but the reduction of the weights of the connections to the points located closer leads to the reduction of the current value of $\boldsymbol{ZG_p(\Delta t_s, p)}$. At a value of **p** = 15 for the ADD procedure, the inequality (2.15) is satisfied. The minimization criterion is reached and at p = 15 the number, positions of the servicing points and the association matrix $[a_{ij}]$ for the next slot can be determined.

The following conclusions can be made while studying the Objective Function behavior, the progress of the optimization procedure and the properties of the ADD/DROP heuristics:

- the algorithmic procedure based on the empirical rule for the "preferred choice for service" reduces the probability of non-prospective candidates to become members of the median set;
- the competitive choice of a successive member of the median set through trial and error of $[N\text{-}p(x_j)]$ possible variations for choosing a "new member" ensures the prospective for reaching the most favorable current set of servicing vertices for the value of $p(x_j)$ at each step of the algorithm.

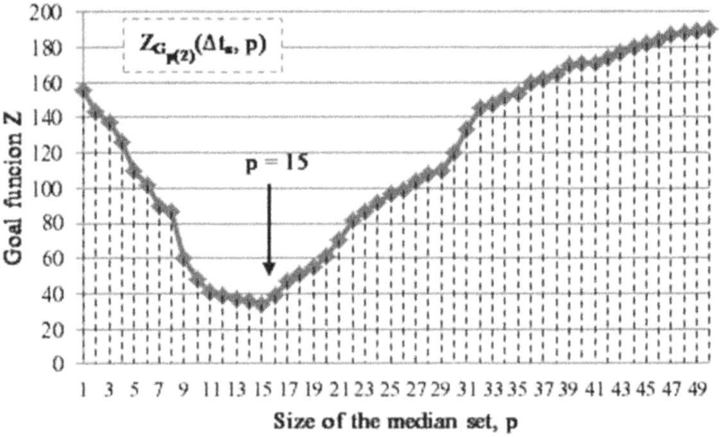

Figure 5.3 ADD/DROP Heuristics Performance Results When Solving the Problem of Finding a P-Median in a Weighted Graph

The convergence of the optimization procedure is controlled at each step of the algorithm through inequalities (2.15) and (2.18). If by adding and/or reducing of a term of the median set both inequalities are not satisfied simultaneously, there is a mistake at the stage of Objective Function modeling and compiling. I this case correction procedures should be applied, related to increasing the individual attraction or individual saturation rates for specific medians in one or more subsets.

5.6 Example of Application of ADD/DROP Heuristics for Solving the Cascade Problem of Hierarchically Connected Sets

For majority of SSs with PCR the application of the model and the related optimization procedure for finding the dimension of the median set leads to a complementary solution concerning the system architecture depending on the service provision mechanism. An example is the development of an efficient model for the provision of effective emergency medical care in a big city. Emergency medical care belongs to the class of SS with PCR due to the fact that the patient should not be denied when requesting a service, i.e. the service is without blocking and with a pro-active control regarding the number and places of the mobile units (ambulances). A hypothesis exists on the parameter which characterizes the period a unit reaches the patient. It can be quite long in case of large distance from the active treatment center. It means that then the risk for the patient's health increases. At the same time the Objective Function of this SS should minimize the risk for all potential patients' health and life, regardless of how many of them and at what address they would need an emergency medical care.

The ideal architecture of an emergency medical care system in a big city supposes several levels of treatment:

first level: mobile units;

second level: intermediate general medical centers for reanimation and primary diagnostics;

third level: specialized centers for active treatment.

The modeling of such a SS involves the definition of values for the key properties (key performance indicators) of the centers at each of the servicing levels and the definition of a general Objective Function related to the multilevel architecture of the system in accordance with:

the slot-based association mechanism of asynchronously incoming user requests;

the model for the characterization of the non-blocking SEs service properties;

the algorithm for the dynamic estimation of the number, place and properties of the active SEs.

Figure 5.4 shows the result of the implementation of the model for the characterization of the servicing properties of non-blocking SEs in an emergency care system. The system has a randomly generated multilevel architecture and a distribution of cases in a single slot.

When forming the model it is assumed that

- The system is developed in 3 servicing levels

  Level "1"- Center SE
  Level "1.N"- Intermediate SE
  Level "1.N.M"- Mobile SE

- The saturation rate of the SEs of level "1.N.M" is assumed to be $\boldsymbol{Ps}_{1.N.M} = \mathbf{3}$
- The attraction rate of the SEs of level „1.N.M" -$\boldsymbol{Pa}_{1.N.M}$ is presented as **the** service zone $\boldsymbol{Pa}_{1.N.M} \approx$ "Level 1.N.M" located on the map of the settlement (Figure 5.4)
- The saturation rate of the SEs of level "1.N" is assumed $\boldsymbol{Ps}_{1.N} = \mathbf{5}$

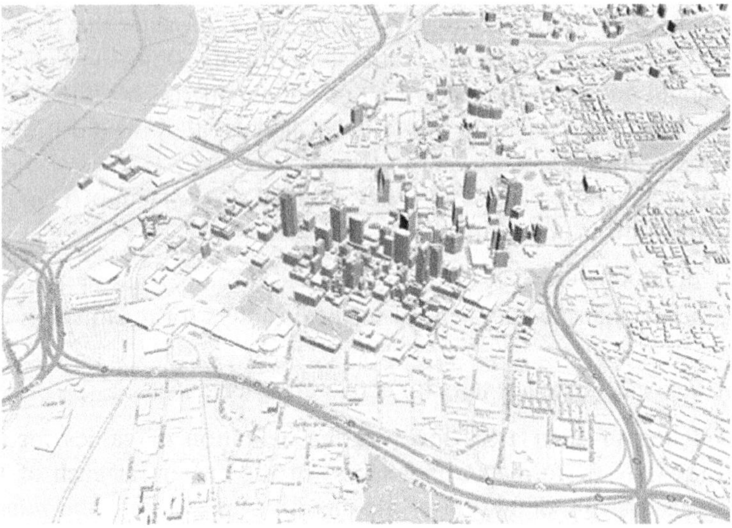

Figure 5.4 Model of a mass service system without blocking and with an emerging necessity for a pro-active control

- The attraction rate of the SEs of level "1.N" -$\boldsymbol{Pa}_{1.N}$ is presented as service zone $\boldsymbol{Pa}_{1.N} \approx$ "Level 1.N" positioned on the map of the settlement (Figure 5.4)
- For Level "1" it is assumed that the saturation rate is $\boldsymbol{Ps}_{1.N} > 5$, which is enough for the center which consolidates and manages the SS resources. On basis of the same logic it is assumed that the attraction rate for the whole urban development territory on which the functionality of the system is defined equals $\boldsymbol{Pa}_1$, i.e. provision of "on-time" and effective medical care.

The evaluation of the applicability of the ADD/DROP heuristics for this example is aimed on the one hand at the necessity of an application adaptation of the algorithm and on the other to "visualize" the specifics of the proactive control as an approach for solving the problem of SEs blocking in SS with PCR.

The system shown on Figure 5.4 is highly sensitive to the blocking of the SEs. The system has its own quasi-stationary structure determined by the place for the physical disposition of the intermediary centers – Level "1.N"and Level "1". Subject to pro-active control for this example is only the SEs of Level "1.N.M".

From this analysis one of the key practical rules for the application of the dynamic proactive control is found: *definition of the architectural levels (layers) for which pro-active management in its entire functionality is applicable*. Other important elements in the preliminary analysis for the application of the dynamic pro-active control with ADD/DROP heuristic approach is the *definition of the control levels in the architecture (3 levels in the example shown in* Figure 5.4*) and the properties of the SEs for each of these levels and the adaptation of the proactive control mechanism to the process of dynamic vertical generation of the current service architecture*. The dynamic vertical (from the lower to the higher architectural levels) generation of the service architecture is implemented practically by means of recursive addressing in a planar ADD/DROP heuristics procedure for determining *the number, position and map of servicing requests association* for each separate SE of the specific architectural level.

A basic problem of the model application is the definition of the number of SEs at each level of the multi-level service architecture. The problem is objectively driven by the fact that SEs (their number respectively) play a key role in solving the optimization problem, i.e. finding the Objective Function's minimum (equations (2.6) and (2.8)). From an optimization point of view the physical meaning of the SEs is the participation of their price F_j as an argument in the Objective function (2.6). In this sense the optimal number of SEs is connected to the minimum price of the weight connectivity, between the SEs and the users (or service request), and with the minimum total price of the SEs themselves as well. A conclusion is that two approaches are applied for determining the number of SEs. The first approach consists of determining their number in advance and the planar procedure determines only the location of the SEs [16–20]. The second approach is related to the simultaneous determination of the number and the distribution of the SEs during the implementation of the planar procedure. The optimal number of SEs, however, does not always coincide with the necessary number of SEs to avoid the situation in which, after solving the blocking collisions, non-associated (i.e. not serviced) requests and/or users remain.

The non-associated requests problem can be solved by the introduction of the so-called correcting procedure which aims at adapting the result for problem solving on finding the p-medians in a weighted graph to the specifics of a real SS.

In order to present the essence of the so-called correcting procedure two objective functions have to be introduced - leveled and generalized. Next equation

$$Z_G^k(\Delta t_s) = \sum_{i=1}^{N^k} \sum_{j \in \overline{X_p}^k(\Delta t_s)}^{p^k(\Delta t_s)} a^k(\Delta t_s)_{ij} w_{ij}^k +$$

$$+ \sum_{j \in \overline{X_p}^k(\Delta t_s)}^{p^k(\Delta t_s)} a^k(\Delta t_s)_{jj} \left[F_j^k + Fd_j^k \left(\sum_i^{N^k} a^k(\Delta t_s)_{ij} w_{ij}^k \overline{S^k(\Delta t_s)_i} \right) \right] \tag{5.20}$$

represents the level objective function, a result of expression (2.8) by introducing the level index of architecture "k". Building up the system's architecture vertically is achieved via:

$$\begin{aligned} &X^{(1)} \to \overline{X_p}^{k=1} \\ &\overline{X_p}^{k=1} = X^{(2)} \to \overline{X_p}^{k=2} \\ &\ldots \\ &\overline{X_p}^{k-1} = X^{(k)} \to \overline{X_p}^{k} \end{aligned} \tag{5.21}$$

Every following level is formed by the "lower" level, the set of nodes of the weight graph at the k-th level $X^{(k)}$being assumed to be the median set$\overline{X_p}^{k-1}$ determined by the optimization procedure for level *k* - 1. Thus, the generalized Objective Function for slot $\Delta t_s$is established as the sum of the objective functions of each of the "generation" levels of the service system architecture, that is:

$$Z_G(\Delta t_s) = \sum_{k=1}^{L} Z_G^k(\Delta t_s). \tag{5.22}$$

With the dynamic pro-active control it is essential that the Objective Function should be minimized for each slot Δt_s by recursively addressing the planar ADD/DROP-heuristics procedure. The dynamic vertical recursive "generation" process aims at establishing the current servicing architecture of the SS. In order to achieve the application functionality of the specific SS no non-associated objects should remain at any of the architecture's levels. Should non-associated requests and/or intermediate SEs remain on at least one of the levels *k* during the implementation of the model and of the optimization procedure the so-called correcting procedure will be activated.

The corrective procedure for the k level involves the execution of the planar association procedure (the ADD/DROP procedure) once again with the implementation of two correcting policies:

- *Correction policy 1*. Increasing the dimension of the median set by 1
  Step 1: Adding a new term to the median set

$$\left|\overline{X_p}^k\right| = \left|\overline{X_p}^k\right| + 1 \tag{5.23}$$

  Step 2: By executing the ADD/DROP heuristics procedure with the new dimension of the median set the p+1 term of the median set is determined and the association matrix is formulated.

  Step 3: Verification is performed whether conditions (2.2) and (2.3) are satisfied for exceeding the saturation rate and/or the attraction rate respectively for a certain median node (servicing element).

  Step *4:* If all requests are already associated, the formulation of $X^{(k+1)} = \overline{X_{|p+1|}}^k$ is made and the new servicing architecture of all levels above the k one is generated. If non-associated requests are available switching back to step 1 is performed.

- *Correction policy* 2. Changing a property of a SE
  This correcting policy influences the properties of the SEs which are closest "by weight" in relation to the non-associated requests and/or users:
  Step 1: Determining the subset$\overline{X_c}^k \in \overline{X_p}^k$, with SEs $x_i{}^k$, which are closest "by weight" in relation to the non-associated requests and/or users. For every SE$x_i{}^k$is applied:

  Step 2: If $\boldsymbol{Ps}_k(xc_i{}^k)$ $=\boldsymbol{Ps}_k$ + $\boldsymbol{n}_a$, where $\boldsymbol{n}_a$is the number of the non-associated users which have the potential to be serviced by $xc_i{}^k$, a functionality is added to increase $\boldsymbol{Pa}_k(xc_i{}^k)$ until association of all $\boldsymbol{n}_a$candidates is reached.

  *Step 3:*If $\boldsymbol{Ps}_k(xc_i{}^k)$ $>\boldsymbol{Ps}_k$ + $\boldsymbol{n}_a$, a functionality is added to increase $\boldsymbol{Ps}_k(xc_i{}^k)$ and switching to Step 2 is then needed.

Correction policy 2 practically does not require the generation of a new architecture at the higher levels when application of only step 2 of the policy is sufficient. When application of Step 3 is also necessary – then switching is made to the generation of the new servicing architecture at all levels higher than the k-th one. This is necessary, because the maximum volume of the requests serviced in a servicing center at the lower level is changed.

It then requires a revaluation of the capacity of the higher levels with the aim of preventing a potential blocking in a SE at the higher architectural levels.

5.7 Conclusion

The pro-active model for dynamic control of service systems without blocking is applicable for the solution of a wide range of technological problems. The modeling of cascade connected service points systems by the application of an ordered pair of properties – attraction rate and saturation rate provides the possibility of solving three classes of problems concerning the architecture of SS with PCR:

Problem class 1: Designing the architecture of newly created SS without blocking.

- Developing a wide range of architectural models by the application of SEs with different properties.
- Testing the architectural model versions by simulation with different geographic density and intensity of the servicing requests.
- Evaluation of the cost of acquisition and operation of future architectural models.

Problem class 2: Analysis and optimization of operating service systems architectures.

- Evaluation of the efficiency of the current topology and properties of the SEs.
- Analysis of the necessity of introduction and location of SEswith basic properties different from the current ones.
- Simulation and testing of scenarios for the optimization of the costs for acquisition of new servicing centers and for the development of existing ones.

Problem class 3: Optimization of existing architectures by the introduction of intermediary levels of virtual SEs without blocking and the transformation of the system from a class with blocking into a class without blocking of the SEs. This problem is a version of the class 2 problem. This type of transformations is typical for continuous service cycle systems. For these systems the establishing of virtualized models and the transformation from a servicing with blocking to servicing without blocking must be performed without termination of the main servicing cycle.

In addition to the 3 classes of application problems the pro-active control model is also applicable to the development of emergency servicing schemes – simulation and evaluation of the efficient servicing plans when one or more servicing elements of the system drop out.

The developed model, based on the definition of two basic properties of the SEs and the servicing, requests a dynamic association mechanism integrated in a PCR scheme. The model presents a complete apparatus for the design and analysis of architectural models of SS without blocking of the SEs.

To this class of systems belong:

- the greater part of the communal services systems – power distribution network, gas distribution network, waste collection, etc.
- cellular mobile radio networks.

A basic advantage of the PRC model is its universality as an instrument for modeling, studying and analysis in view of the system's inertness as well as the character and/or type of service provided.

The PCR model is based on a slot principle of forming the control impact (command), the regulation of the absolute slot duration Δt_s being a function of the system's dynamics in relation to the changes in the geographic density and the intensity of service requests. The fact that the algorithm for forming the control impact requires a minimum resource of time for a solution, a possibility for a sufficient reduction of Δt_s is provided. This is possible up to the moment when regardless of the system's dynamics the statistical forecasting from the current slot database will provide a sufficiently realistic picture of the expected density and intensity of requests for the next slot, within which the control impact will be applied.

Major characteristic feature of the control impact formation is that it is realized as a system function, i.e. it has an effect on the overall system architecture and it is based on the idea of achieving a topological adequacy of the service while minimizing the service costs.

In relation with the applicability the model with generalization of the properties of the SEs has the following advantages:

- The properties of attraction rate and saturation rate can be measured authentically and they can be practically attributed to every SE regardless of the application field as these two properties are typical for the service provision process from application viewpoint.

- The properties of the SEs participate as parameters in the composition and calculation of the system objective function. The resulting weighted values, after the association, are representative in view of the evaluation of the influence of each ordered couple of properties of a SE type on the total weighted cost of service of the system as a whole.
- A consequence of the model is the control impact, a result of an optimization problem for linear programming, for which heuristic algorithms with low computational complexity and relatively fast convergence are applicable.
- The character of the control impact based on an association matrix and the related weighted evaluations of the individual, group and overall system weighted costs of service provisioning ensure its effective reversive application, back from the model to the object to be controlled.
- The model is reversible as an application concerning the object. It is possible to be applied as a future service system architecture (direction from model to object) or for the analysis and optimization of an existing servicing system (direction object-model-object).

From the practical perspective of application of the model in computing control devices the requirements to the hardware configuration (memory for constant and variable computations) are minimal. A limited set of instructions is sufficient for the execution of the ADD/DROP heuristics because all calculations are scalar. When the control mechanism operates with one current and one predicted time slot the Association Database volume is minimal and with a dimension determined by the number of recordings one for each active user. Each of the recordings consists of two elements: geographic location of the user (X_i,Y_i) and an element for the properties description of the currently requested services set.

The model presented in this chapter for SS with PCR with application of ADD/DROP heuristics for association of the servicing requests is applicable to a broad class of systems without blocking, where the control quality is context-independent of the application field.

The migration towards SS with PCR does not imply substantial architectural and functional changes in the system and affects only the communication model for the provision of the management process. The restrictive condition for the application is related to the maintenance of technical resources for the dynamic distribution of the valid pro-active control impact to the SEs and to the users when passing to the next servicing slot.

References

[1] H. Kobayashi, A. G. Konheim. Queueing Models for Computer Communications System Analysis. IEEE Transactions on Communication, 25(1):2–29, January 1977.

[2] F.B. Nilsen. Queuing systems: Modeling and simulation, analysis. Research Report 259 University of Oslo Department of Informatics, Oslo, ISBN 82-7368-185-8 ISSN 0806-3036, April 1998.

[3] H. Javanshir, M. Jokar, M. Hadadi, M. Zakerinia. Analysis Of Non-Standard Queue Systems By Using A Hybrid Model. In Proceedings of International Conference on Computers & Industrial Engineering, Los Angeles CA, USA, 459–464, October 23–26, 2011:.

[4] P. Tsigas, Yi Zhang. A Simple Fast and Scalable Non-Blocking Concurrent FIFO Queue for Shared Memory Multiprocessor Systems. In Proceedings of ACM Symposium on Parallel Algorithms and Architectures, Crete Island, Greece, :1–22, July 4–6, 2001.

[5] T. Nilsson, G. Wikstrand, J. Eriksson. A collision detection method for multicast transmissions in CSMA/CA networks. Wireless Communications and Mobile Computing, 7(6):795–808, August 2007.

[6] N. Chirdchoo, W.-S. Soh, K. C. Chua. Aloha-Based MAC Protocols with Collision Avoidance for Underwater Acoustic Networks. In Proceedings of INFOCOM 2007. 26th IEEE International Conference on Computer Communications. IEEE, Anchorage, AK, :2271-22756-12 May 2007.

[7] European Commission. (2010, August) European Programme for Critical Infrastructure Protection.

[8] J. Reese. Methods for Solving the p-Median Problem: An Annotated Bibliography. Mathematics Faculty Research, Trinity University, Trinity, Paper 28, 2005.

[9] M.J. Varnamkhasti. Overview of the Algorithms for Solving the P-Median Facility Location Problems. Advanced Studies in Biology, 4(2):49–55, 2012.

[10] J. E. Beasley. A note on solving large p-median problems. European Journal of Operational Research, Elsevier, 21(2):270–273, August 1985.

[11] A. Ceselli. Two exact algorithms for the capacitated p-median problem. Quarterly Journal of the Belgian, French and Italian Operations Research Societies, Springer, 1(4):319–340, December 2003.

[12] M. Charikar, S. Guha, ´E. Tardos, D. B. Shmoys. A constant-factor approximation algorithm for the k-median problem. In Proceedings of

the 31st Annual ACM Symposium on Theory of Computing, Atlanta, Georgia, USA, :1–10May 1–4, 1999.

[13] J. Cheriyan , R. Ravi. Approximation Algorithms for Network Problems. Lecture Notes. University of Waterloo, Carnegie Mellon University, 1998.

[14] Yu-Chiun Chioua, Lawrence W. Lan. Genetic clustering algorithms. European Journal of Operational Research, Elsevier, 135(2):413–427, December 2001.

[15] N. Christofides. Graph Theory: An Algorithmic Approach. New York, USA, Academic Press Inc, 1975.

[16] R.D. Galvão. A Dual-Bounded Algorithm for the p-Median Problem. Operations Research, Informs, 28(5):1112–1121, September–October 1980.

[17] R.D. Galvão. A graph theoretical bound for the p-median problem. European Journal of Operational Research, Elsevier, 6(2):162–165, February 1981.

[18] R. L. Rardin, R. Uzsoy. Experimental Evaluation of Heuristic Optimization Algorithms: A Tutorial. Journal of Heuristics, Springer, 7(3):261–304, May 2001.

[19] L. A. N. Lorena, J. C. Furtado. Constructive Genetic Algorithm for Clustering Problems. Evolutionary Computation, MIT Press Journals, 9(3):309–327, Fall 2001.

[20] K. Jain, V. V. Vazirani. Approximation algorithms for metric facility location and k-Median problems using the primal-dual schema and Lagrangian relaxation. Journal of the ACM, 48(2):274–296, March 2001.

Biographies

Associated Professor Oleg Asenov, has received his MSc degree at the Technical University of Gabrovo and PhD degree at the Technical University of Sofia. He has more than 20 years of teaching and research experience in the field of Telecommunications. The major fields of scientific interest are modeling, simulation and design of computer networks based on graph theory and applied heuristics algorithms. Currently he is Associated Professor at the St.Cyril and St.Methodius University of VelikoTyrnovo, Member of IEEE.

Pavlina Kolevareceived M.Eng. and Ph.D. degrees in Telecommunications from the Technical University of Sofia, Bulgaria, in 2001 and 2013, respectively. Since 2001, she has been an Assistant Professor at the Technical University of Sofia. She is involved in projects, related to software design, development and support of various types of systems. Her main research interests are in Information Theory, Communication Networks, Game Theory, Cognitive networks, and Next Generation Networks.

Professor Vladimir Poulkov PhD, has received his MSc and PhD degrees at the Technical University of Sofia. He has more than 30 years of teaching and research experience in the field of telecommunications. The major fields of scientific interest are in the fields of Information Transmission Theory, Modulation and Coding. His has expertize related to interference suppression, power control and resource management for next generation telecommunications networks. Currently he is Dean of the Faculty of Telecommunications at the Technical University of Sofia. Senior Member of IEEE.

6

Machine-to-Machine Communications for CONASENSE

Kwang-Cheng Chen[1] and Shao-Yu Lien[2]

[1]INTEL-NTU Connected Context Computer Center, National Taiwan University, and visiting the Research Laboratory of Electronics, Massachusetts Institute of Technology, Taiwan
[2]Graduate Institute of Communication Engineering, National Taiwan University, Taiwan

Machine-to-machine (M2M) communications emerge to autonomously operate to link interactions between Internet cyber world and physical systems, or between cyber world and social systems. We present the technological scenario of M2M communications consisting of wireless infrastructure to cloud, and machine swarm of tremendous devices. Such devices or machines are capable of sensing the environment or useful information and autonomous transportation of such sensed information to effectively operate physical systems/world. Further practical realizations are explored to complete fundamental understanding and engineering knowledge of this new communication and networking technology front.

6.1 Introduction

Following tremendous deployment of Internet and mobile communications, Internet of Things (IoT) and Cyber-Physical Systems (CPS) emerge as technologies to combine information communication technology (ICT) with our daily life [1 – 3]. By deploying great amount of machines that are typically wireless devices, such as sensors, we expect to advance human being's life in a significant way. In particular, autonomous communications among machines of wireless communication capability creates a new frontier of wireless communications and networks [4 – 5]. In this chapter, we will survey

Convergence of Communications, Navigation, Sensing and Services, 127–180.

some technological milestones and research opportunities toward achieving machine-to-machine (M2M) wireless communication ultimately serving human beings, through sensing, communications, and control cyber-physical systems.

Figure 6.1 delineates the fundamental network architecture of cloud-based M2M communications, consisting of cloud, infrastructure, and machine swarm (or machine oceans, to stand for a great amount of machines). Networking in the cloud, typically done by high-speed wired/optical networking mechanism, connects data centers, servers for applications and services, and gateways to/from the cloud. The infrastructure interconnects cloud and machine swarm/ocean, which can be wired or wireless. In this paper, we focus on wireless infrastructure, which allows flexibility and mobility to enable M2M applications and services. For potentially wide geographical range and diversity of deployment, cellular systems play the key role in (wireless) infrastructure. We therefore introduce 3GPP type of systems supporting M2M [5–7] in details. The data aggregators (DA) are transmitting/receiving, collecting, or fusing information between infrastructure and machine swarm, which can be considered as the access points to infrastructure networks. Finally, the number of machines can go up to trillions according to various reports. Such a huge number of wireless devices form machine swarm or machine ocean, and

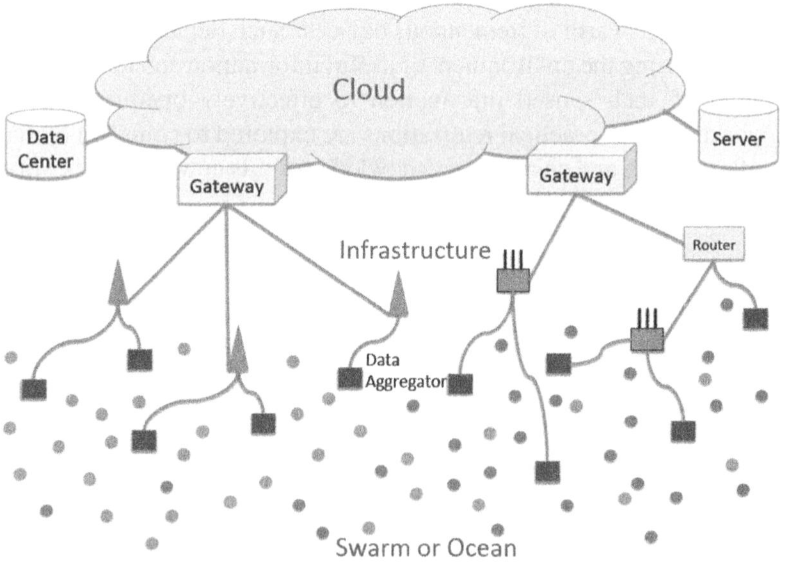

Figure 6.1 Cloud-Based M2M Communications

create a new dimensional technology challenge in wireless communications and networks, after the triumphs of wireless personal communications for billions of handsets in past 2 decades. It also suggests potential challenges in deployment, operation, and security and privacy.

Consequently, the organization of this chapter surveys and highlights technology for M2M wireless communications as follows. Section 2 is dedicated to wireless infrastructure. Section 3 summarizes technology to achieve efficient communications in machine swarm/ocean, particularly under spectrum sharing scenario. Section 3.1 orients sensing methodologies by minimizing communication overhead. Various issues in deployment, operation, and security and privacy, are explored in Section 4.

6.2 Wireless Infrastructure

To practice M2M communications, few realizations of M2M communications have been proposed, such as leveraging Bluetooth (IEEE 802.15.1), Zigbee (IEEE 802.15.4), or WiFi (IEEE 802.11b) technologies. However, there is still no consensus on the network architecture for M2M communications over these wireless technologies. Considering that the ultimate goal of M2M communications is to construct comprehensive connections among all machines distributed over an extensive coverage area, the network architecture of M2M communications leveraging these wireless technologies can generally be considered as the heterogeneous mobile ad hoc network (HetMANET), and faces similar challenges that can be encountered in the HetMANET. Although a considerable amount of research has provided solutions for the HetMANET (connections, routing, congestion control, energy-efficient transmission, etc.), it is still not clear whether these sophisticated solutions can be applied to M2M communications due to constraints on the hardware complexity of a MTC device. Because of these potential concerns, scenarios defined by 3GPP thus emerge as the most promising solution to enable wireless infrastructure of M2M communications [5][8][9] useful for CONASENSE systems.

6.2.1 Ubiquitous Connections via 3GPP Heterogeneous Network (HetNet) Architecture

To provide ubiquitous wireless connections for user equipments (UEs) of human-to-human (H2H) communications in indoor and outdoor environments, a special network architecture known as heterogeneous network (HetNet) is introduced by 3GPP LTE-Advanced [10 – 12]. In the HetNet,

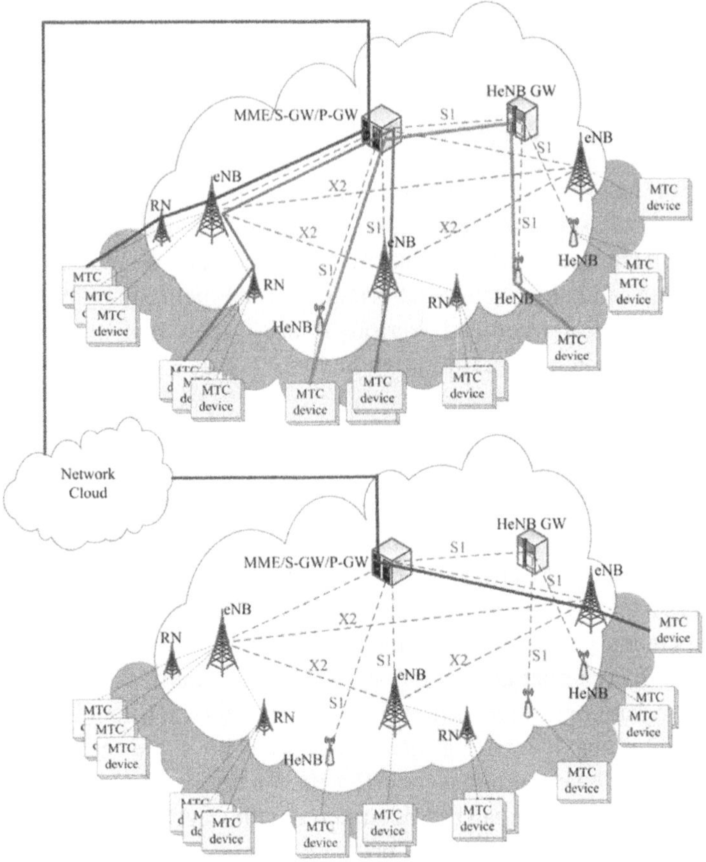

Figure 6.2 Connection Scenario of 3GPP MTC Devices (a Similar Illustration is Shown in [11]).

in addition to conventional macrocells formed by evolved universal terrestrial radio access (E-UTRA) NodeBs (eNBs), there are picocells formed by small transmission power eNBs deployed underlay macrocells to share traffic loads of macrocells, femtocells formed by HeNBs deployed underlay macrocells to enhance signal strength and coverage in the indoor environment, and relay nodes (RNs) deployed in coverage edges of macrocells. The 3GPP infrastructure provides higher layers connections among all stations of eNBs, HeNBs, and RNs. Although, in the HetNet, there is potential interference between small cells in the air interface (of picocells, femtocells, and RNs) and macrocells, such interference can be effectively mitigated by applying recent solutions ([15] for picocells, [14 – 18] for femtocells, and [19][20] for RNs). As a consequence, by attaching to these stations, ubiquitous connections among all machines can be provided. In 3GPP, a machine is referred to a machine-type communication (MTC) device. An illustration of the M2M communications in 3GPP is shown in Figure X. By the 3GPP infrastructure, a secure, energy efficient, reliable and mobility-empowered connection at the same level of common UEs can be provided for M2M communications.

Although 3GPP provides all these technical merits, it does not suggest a successful practice of M2M communications. The most challenging issues lie in a severe signaling congestion on the air interface and a complicated management in the network, as it is estimated that the number of MTC devices will be 1000 times larger than the number of UEs [21]. Although considerable solutions have been proposed for solving these critical issues by cooperative among stations [22 – 25], a group based operation of MTC devices has been regarded as a promising direction [8][26 – 28], and device-to-device communications later.

6.2.2 D2D Empowered Group Based Operations of MTC Devices

The primary idea of grouping a number of MTC devices into a swarm is to reduce the number of communications between a MTC device and 3GPP E-UTRA and evolved packet core (EPC). That is, a group header gathers requests, uplink data packets and status information from MTC devices in the group, and then forwards such traffic to a station of 3GPP. The group header also relays management messages and downlink data packets from a station of 3GPP to MTC devices in the group. By avoiding direction communications between a MTC device and 3GPP E-UTRA, signaling congestion on the air interface and a complicated management in the network can be alleviated. This is the spirit of the group-based management defined by 3GPP [8]. For M2M communications, MTC devices can be grouped (i) *logically based on service demand patterns of MTC devices*, or (ii) *physically based on locations of MTC devices.*

For (i), considering one of major applications of MTC communications is to collect measurement data from MTC devices (e.g., data reports from meters in smart grid networks or navigation signals from positioning sensors in navigation networks), traffic of such application typically has characteristics of periodical packets arrival, small data (each MTC device only sends or receives a small amount of data), and with certain hard or soft jitter constraints. The schedule of a large amount of packets with small data to meet corresponding jitter constraints is a huge computational burden and challenge. To tackle this challenge, MTC devices of similar characteristics can be merged into a group *logically*, then resources for these MTC devices can be scheduled and managed in the basis of groups. Consider M groups of MTC devices indexed by i=1,...,M. The packet arrival period of MTC devices in the ith group is $1/\lambda_i$, and a granted time interval with length τ is allocated to each group periodically (based on the packet arrival period of the group) for packet transmissions of MTC devices in the group. When granted time intervals are allocated according to present priority among groups and a group with a larger λ_i has a higher priority, the jitter of packets in the ith group is bounded above by [26]

$$\delta_i^* = \tau + \sum\nolimits_{k=1}^{i} \left\lceil \frac{\gamma_k}{\gamma_i} \right\rceil \tau, \; for \; i = 2, \ldots, M \tag{6.1}$$

if $\delta_i^* + \tau < 1/\gamma_i$, and $\delta_i^* = \tau$ for i=1. This result significantly facilitates to reduce the complexity on scheduling. Denote δ_i as the jitter constraint of packets in the ith group. The scheduler only needs to check $\delta_i \leq \delta_i^*$ for all i, then it is guaranteed that jitter constraints of all packets in all groups can be satisfied.

For (ii) to *physically* group MTC devices, a new communications paradigm is essential: a direct communication between a group header and an MTC devices. To enable such a direction communications among MTC devices, we shall note a communications paradigm that will be defined in 3GPP Rel. 12 referred to device-to-device (D2D) communications [29 – 31]. To significantly reduce the transmission energy at the UE side, always transmitting packets to E-UTRA then E-UTRA relaying these packets to the destination UE may not be an optimum scheme, especially when the source UE is located nearby the destination UE. To orientate this situation, 3GPP plans to provide the interface and protocols for direct packets exchanges among UEs as D2D communications. By facilitations of the interface and protocols of D2D communications, MTC devices in a group can communicate with the group header without the intermediate of 3GPP E-UTRA to enable the group-based operation of MTC devices. Precise system design and performance evaluations are still subject to further study.

6.2.3 Cognitive Operations of MTC Devices

Even though the group based operation of MTC devices can potentially alleviate signaling congestion and management burden in M2M communications, these issues may not be resolved when the number of MTC devices grows enormously in the future. Under this circumstance, the operator may need to deploy more stations of E-UTRA and separate traffic as well as the management of UEs and MTC devices. That is, there can be E-UTRA stations for UEs and E-UTRA stations for MTC devices coexisting with each other, as shown in Figure 6.2, to relax signaling congestion and management burden. However, under such a coexisting deployment, interference between conventional human-to-human (H2H) communications (that is, conventional links between UEs and E-UTRA) and M2M communications turns to be a challenging issue. Although interface of E-UTRA (i.e., X2 interface) can leverage a centralized coordination for interference mitigations, this scheme is not suggested, due to a centralized coordination creating significant management burden. An appropriate solution for interference mitigation between H2H and M2M communications lies in a distributive resource management, and a powerful technology known as *cognitive radio* (CR) is particularly noted.

Since E-UTRA stations for UEs and E-UTRA stations for MTC devices belong to identical technology, there is no priority among these stations. This operation is exactly a particular mode of the CR operation referred as the "interweave" coexistence [32]. The operation of the interweave coexistence between H2H and M2M communications can be outlined in Figure 6.3 [33].

MTC devices only utilize unoccupied radio resources from that of H2H communications. For this purpose, MTC devices perform interference measurement and report measurement result (indicating occupied resources by H2H communications) to the group header by D2D communications technology. As a result, the group header shall acquire interference situations of all MTC devices in the group. However, if all MTC devices shall report the measurement result to the group header, the channel to the group header suffers severe congestions. To tackle this issue, a powerful technology known as *compressed sensing* is particularly noted. Compressed sensing origins as a signal processing technology, which is able to sample a (audio/image) signal with a sampling rate far lower than that of the Nyquist rate and the sampled signal can be recovered with an acceptable error rate if certain constraints can be satisfied. Specifically, denote the number of MTC devices in a group as V. At each measurement, if each MTC device only perform interference measurement with a probability q, then the expected number of MTC devices involved in interference measurement and reporting result is $Vq \leq V$. For this purpose, the coverage of a group is divided into N isotropic grids.Denote

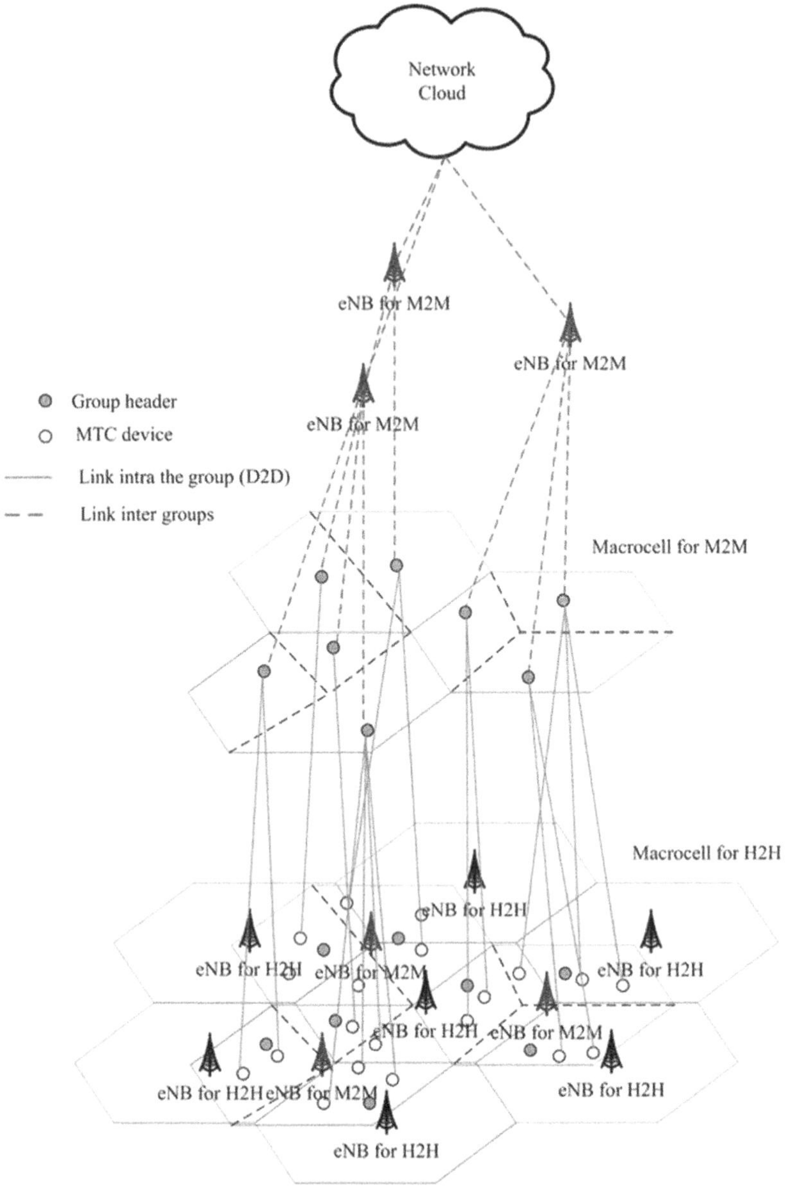

Figure 6.3 Coexistence of E-UTRA for M2M Communications and H2H Communications (a Similar Illustration is Shown in [33])

the true interference from H2H communications as $\Psi = [\varphi_1\varphi_2 \ldots \varphi_N]^T$, the compressed sensing is obtained by multiplying a sample matrix on Ψ as

$$y = \Phi\Psi + \varepsilon, \tag{6.2}$$

where Φ is a $R{\times}N$ matrix with each element taking "1" with probability qV_n/V, and taking "0" with probability of 1- qV_n/V, where V_n is the number of MTC devices within the *n*th grid in a group. After obtaining *Vq* measurement reports, interference from H2H communications can be reconstructed by searching the minimum l_1 norm of Ψ,

$$\Psi^{*}{=}\arg\min \|\Psi\|_1 \text{ s.t. } \|\Phi\Psi{-}y\|_2 \leq \varepsilon \tag{6.3}$$

through applying the second order corn programming if $R{=}O(K\log\frac{NM}{K})$, where l_p norm of Ψ is

$$\|\Psi\|_p \equiv \left(\sum\nolimits_{n=1}^{N} |\varphi_n|^p\right)^{1/p}, \tag{6.4}$$

K is the sparsity of Φ, $\varepsilon = \|\Omega\varepsilon\|_2$ and Ω is a random basis.

The group header delivers measurement results to E-UTRA stations for M2M communications. E-UTRA stations then perform compressed sensing calculation for the interference (from H2H communications) reconstruction and allocate unoccupied radio resources to the group. MTC devices then can communicate with each other in the D2D fashion in the group, or communicate with MTC devices in other groups by relays of the group header and E-UTRA stations.

Once a radio resource is occupied by M2M communications, this radio resource is regarded as suffering server interference and will not be utilized by H2H communications, as UEs perform channel estimation before transmissions.

By this particular mode of CR operations and the group based operation, challenging issues of signaling congestion and heavy management burden can be effectively resolved.

6.2.4 The QoS Guaranteed Optimal Control for Cognitive Operations of MTC Device

The major concern in CR operations stated above is potential efficiency of radio resources. To avoid interference to/from H2H communications, E-UTRA stations shall allocate radio resources for MTC devices to perform interference measurement (although the number of MTC devices involved in interference measurement can be significantly reduced by the compressed sensing technology mentioned above). These radio resources for interference measurement are overheads, as all MTC devices can not perform data transmissions nor receptions by these radio resources for interference measurement. If radio resources for interference are allocated very frequently, although time-varying interference from H2H communications can be accurately estimated, there is a severe resources waste. On the other hand, if radio resources for interference are allocated very frequently rarely, although overhead is reduced, interference from H2H communications may not be alleviated. As a result, the measurement period can be an very critical factor impacting the performance of MTC devices, especially the most critical QoS guarantees.

To solve this challenge, we particular note an equilibrium of statistical delay guarantees that

$$\Pr\{delay>d_{max}\} \approx e^{-\theta\delta d_{max}} \tag{6.5}$$

as providing a deterministic delay guarantee of $\Pr\{delay>d_{max}\}=0$ over wireless channel has been shown impossible [34], where d_{max} is the delay bound and δ is jointly determined by the arrival process of traffic and the service process of the system. From (6.5), it can be observed that a small θ implies that the system can only support a loose QoS requirement, while a large θ means that a strength QoS requirement can be supported by the system. To reach the equilibrium of (6.5), the effective bandwidth and the effective capacity serve significant foundations.

The effective bandwidth [35], denoted by $\mathrm{E}_B(\theta)$, specifies the *minimum service rate needed to serve the given arrival process subject to a given* θ. On the other hand, the effective capacity, denoted by $\mathrm{E}_C(\theta)$, is the duality of the effective bandwidth, which specifies the *maximum arrival rate that can be supported by the system subject to a given* θ. If θ^* can be found as the solution of $E_B(\theta^*) = E_C(\theta^*)\delta$ can be obtained by

$$\delta = E_B(\theta^*) = E_C(\theta^*). \tag{6.6}$$

Consequently, the system can achieve the equilibrium of the statistical delay guarantee

$$\Pr\{delay>d_{max}\} \approx e^{-\theta^*\delta d_{max}}. \tag{6.7}$$

With the facilitation of above foundations, the optimal control of the measurement period and radio resources allocation for a group of MTC devices can be summarized in the following.

1) Calculates the effective bandwidth $E_B(\theta)$ of the real-time traffic for a group of MTC devices.
2) Set the measurement period to an initial value.
3) Allocate $s = 1$ radio resource to a group of MTC devices.
4) Find the solution of θ such that $E_B(\theta)$=$E_C(\theta)$= δ.
5) Derive the delay violation probability by $\Pr\{delay{>}d_{max}\} \approx e^{-\theta\delta d_{max}}$.
   a) If $e^{-\theta\delta d_{max}}{>}\upsilon$, where υ is the upper bound of the acceptable QoS violation probability, determine s by

$$min_{1\leq s\leq S}\{s\}, \text{ s.t. } e^{-\theta\delta d_{max}} \leq \upsilon \ . \tag{6.8}$$

   b) If (6.8) is not satisfied, decrease the measurement period by one if the current measurement period value is larger than two and repeat Step 4) and 5) until (6.8) is satisfied.

The above optimal control for QoS guarantees enables reliable transmissions of a massive number of MTC devices via the facilitation of the cognitive radio technology and the group base operations of M2M communications. Consequently, the most critical challenges of signaling congestion, spectrum congestion, and heavy management burden on the air interface of M2M communications can be effectively resolved.

More research opportunities exist in the areas of MTC and D2D communications in 3GPP, potential networking protocols to facilitate direct communications among devices if appropriate, and complete realization under cognitive cellular networks [36]. Actually, D2D introduces a new dimensional challenge of spectrum sharing between heterogeneous networks, D2D and cellular system. Such spectrum sharing can be realized by game theory as a topic of network economy [37] to achieve efficient spectrum utilization for communications. Practical operating algorithms and system design for autonomous information exchange in CONASENSE systems are the next engineering step to develop.

6.3 Statistical Networking in Machine Swarm/Ocean

Technology to connect wireless devices under M2M scenarios has been proposed and developed for years, such as RFID, Bluetooth, Zigbee, and WiFi, corresponding to various on-going or announced IEEE 802 standards. The scope of this survey does not focus on such short-range wireless communication technology. Instead, we assume availability for such physical layer wireless connectivity and communication technology, but focus more on new challenges beyond physical transmission, particularly networking in this section.

In machine (or sensor) swarm/ocean, potentially except data aggregators (DAs), energy-efficiency and therefore low-power transmission becomes a must and no way to compromise, no matter for battery life or energy harvesting operation from nature. From current energy harvesting technologies [38], the subsequent wireless communication techniques must be short-range, or mid-range with extremely low data rates. Under a large number of machines, multi-hop ad hoc networking is evitable in realistic deployment and operation. Furthermore, the operating communication protocols in each machine must be simple and energy-efficient in implementation, as further challenges.

Spectrum is always a critical issue in wireless communications. Under the shortage of wireless spectrum, the spectrum utilization for communications in the machine swarm may most likely fall into 2 categories of spectrum sharing: (i) machines as secondary users to dynamically access the temporarily inactive spectrum of primary users, to form networking, which is known as cognitive radio networks (CRNs) [39 – 43]. (ii) multiple networks in the machine swarm to share a dedicated spectrum, which can be treated as (spectrum sharing) heterogeneous wireless networks. In this section, we focus on state-of-the-art technology to implement multi-hop cognitive radio networks and multi-hop spectrum sharing heterogeneous[1] wireless networks, and associated research opportunities for this wide-open knowledge.

Prior to more technologies in-depth, please consider the fundamental technological challenges in front of spectrum sharing multi-hop networking in the machine swarm:

- Facilitation of spectrum sharing ad hoc networks: The ad hoc networking is already difficult [44], and even much more challenging under spectrum sharing [45], particularly no control channel available in the spectrum sharing machine swarm. Spectrum sharing generally invokes *opportunistic networking* to complicate stochastic analysis and operation of networking.
- Cooperation in sensing, relays, and networks: Cooperative sensing [46 – 48], cooperative relay [49][50], and cooperative networks [51][52], are easy to assume and to show excellent performance in a localized view. However, cooperation suggests some critical assumptions behind the

[1]Heterogeneous wireless networks here mean multiple independent (but may cooperate) wireless networks, and slightly different from heterogeneous networks for multi-tier cellular systems in Section 2, though still consistent in general definition.

scene. For cooperative sensing, the bandwidth to collect information from cooperative sensors may be much greater than the required bandwidth to transmit signal. Cooperative relay is good to enhance performance at link level, but it generates more interference, time to occupy radio spectrum though maybe in smaller geographical range, and further trust and security concerns [53]. Similarly, cooperation among networks implies signaling overhead and cost on network management and is neither obvious nor straightforward from many aspects.

- Large but resource efficient multi-hop networks: With almost all machines (or sensors) are relatively simple in hardware and software, to maintain efficient operation for a large ad hoc network of machines is open to human's engineering knowledge. As a matter of fact, the fundamental performance of wireless ad hoc networks has been a research challenge in very recent years, which hopefully suggests the efficient and saleable design of large multi-hop ad hoc networks [54–56] but still remains open.

We will explore various subjects to provide part of the answer for above fundamental technological and intellectual challenges. From Section 2 and similarly in this section, keen readers might already note that we approach the cross-layer system and network design from radio resource directly by diminishing the role of medium access control [57] for the ease of presenting system architecture. Such a concept via interdisciplinary technology is somewhat similar to so-called *layerless dynamic networks* in [56] and shall be explored further in later of this section. In addition, although there lacks widely accepted models for wireless channels and aggregated activities of high-density machine swarm (i.e. large network), the results of ad hoc cognitive radio networks generally apply in the following. We therefore suggest a revolution paradigm to *statistically* network the machines in the swarm, to avoid the difficulty of tremendous control overhead in end-to-end ad hoc networking.

6.3.1 Sensing Spectrum Opportunities for Dynamic Spectrum Access

In spectrum sharing communication in machine swarm, prior to each transmission, each machine shall sense the channel availability first to avoid potential collision destroying spectrum efficiency [39] by facilitating dynamic spectrum access toward ultimate spectrum efficiency. Conventional spectrum sensing to detect or to sense spectrum hole(s) or white space(s) at link level (i.e. for an opportunistic transmission between a CR-Transmitter and CR-Receiver pair) targets at a single primary system and is used to decide between the two hypotheses in discrete-time statistical signal processing, namely

$$y[n] = \begin{cases} w[n] & \\ hs[n] + w[n] & n = 1, \ldots, N \end{cases} \quad \begin{matrix} H_0 \\ H_1 \end{matrix} \tag{6.9}$$

where $y[n]$ represent observation samples, $s[n]$ and $w[n]$ are contributed from PS signal and noise respectively, and h is the corresponding channel gain observed at the CR transmitter. Numerous techniques to achieve the goal of spectrum sensing have been proposed with more details in [58][59], while the detection and estimation techniques include

- Energy detection
- Matched filtering
- Cyclostationary detection
- Wavelet detection
- Covariance-based detection
- Multiple-antenna
- Sequential hypothesis testing and thus universal source coding

Cooperative sensing [46 – 48] has been introduced to alleviate the hidden terminal problem in sensing. However, most spectrum sensing techniques deal with a single link-level transmission of targeting successful transmission signal based on CR-transmitter's observation on spectrum availability, instead of successful transmission for networking purpose. As networking consisting of a number of time-dynamic links, state-of-the-art sensing techniques designed for networking purpose are summarized as follows:

- Spectrum availability of CR-Receiver: As recalled and indicated in [60], the successful transmission over a CR link relies on spectrum availability not only at CR-Transmitter's location but also at CR-Receiver's location. For networking purpose, we shall understand the availability of multiple links in CRN, and likely associated with location information.
- Statistical inference: One major issue in spectrum sensing is the time-lagged information of spectrum availability at earlier time slot of measurement, rather than the time slot in transmission. That is, we execute spectrum sensing at time $t_{n-L}, \ldots, t_{n-1}, t_n$, however, the CR-transmitter transmits at time t_{n+1} and spectrum availability information up to t_n is not immediately useful, as a major blind side in tremendous spectrum sensing research. Fortunately, this can be easily resolved by

applying statistical inference [60][61]. The most straightforward application of statistical inference is to utilize Laplace formula to predict spectrum availability [60]. Suppose we observe spectrum availability for L periods of time to form a Bernoulli trial random process, indicated by $1\left[n-L\right], \ldots, 1[n-1]$, while 1 stands for spectrum available and 0 stands for spectrum unavailable. The probability of spectrum available at time n is estimated by

$$\hat{1}\left[n\right] = \frac{N+1}{L+2} \tag{6.10}$$

where $N = \sum_{l=1}^{L} 1[n-l]$.

- Radio network tomography: Ideally, to achieve perfect sensing for entire wireless network, we have to know the channel statistics of each link in the network, that is, $h_{i,j}$ between node i and node j, $\forall i, j$. For wired networks, a technology known as Internet tomography [62] is developed based on statistics. However, for radio networks, it is much more challenging until cognitive radio network tomography [61] that leverages the concept like medical tomography to transmit a test radio signal into the network and to statistically infer the radio source in the entire wireless network. Subsequently, spectrum sensing can be generalized into tomography by extending statistical inference, beyond pure physical layer detection and estimation for a specific link. Radio network tomography becomes a sort of statistical measuring, processing, and inferring techniques that provide the parameters and traffic/interference patterns for radio network operations at both link level and network level. Hereafter, we illustrate two cognitive radio network tomography examples at network level, beyond physical layer at link level: (i) multiple system sensing (or identification) (ii) radio resource sensing. For multiple-system sensing, the radio signal received at the CR-Rx can be written into

$$\mathbf{y} = \sum_{i=1}^{Q} I_{active}^{P}\left[i\right] \cdot a_i s_{\mathrm{i}} + w \tag{6.11}$$

where $\mathbf{y} = \left[y\left[1\right], \ldots, y[N]\right]^T$ is the observed signal vector, $a_i s_{\mathrm{i}}$ represents transmitted signal weighted by the complex channel gain, embedded by the noise. $I_{active}^{P}\left[i\right]$ indicates the ith system to be active. We can use different methods to identify active systems among Q candidate systems. For radio resource sensing in a CRN, referring the simple interference model, the optimal

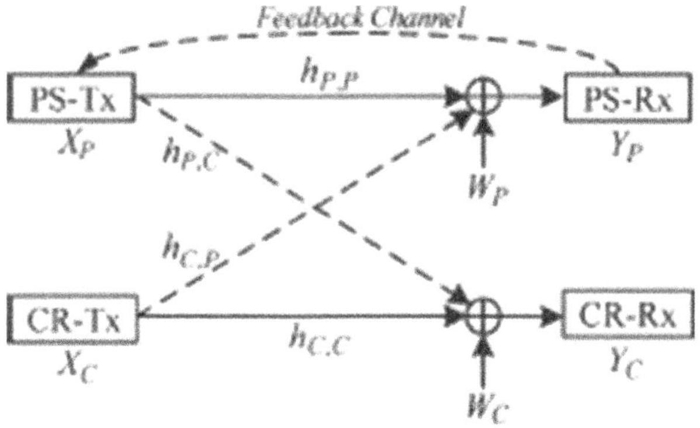

Figure 6.4 Interference Model for a Simple Cognitive Radio Network [61]

transmission power can be determined by sending a probing signal into radio network and observing the adaptive modulation and coding change.

- Spectrum map: A major difference of spectrum sensing between a cognitive radio link and a cognitive radio network lies in the preference of spectrum opportunity (or available radio resource) assessment on the tendency of future hops for cognitive radio network. In other words, *spectrum map* indicating spectrum availability associated with location would be precisely wanted in CRN and any spectrum sharing networks. The challenge here, beyond cooperative sensing, is to collect such spectral-spatial information with minimal overhead and thus bandwidth, preferred without such overhead. Two methods are suggested to accomplish such a goal: *synthetic aperture radar* (SAR) [63] and *compressive sensing* [64]. As a matter of fact, compressive sensing to take advantage of the sparsity nature of spectrum availability for networking (i.e. no need to know the complete spectrum information) could play an interesting role for spectrum sensing [64][65].
- Information fusion: Since spectrum sensing requires *information fusion* from spectrum sensors, it suggests another highly potential alternative to design a spectrum-efficient spectrum sensing for CRN. We can use a single sensor, that is, the transmitter itself, to sense multiple kinds of heterogeneous information from CRN. With a newly developed theory of *heterogeneous information fusion and inference* (HIFI), we can infer the status of CRN spectrum availability as good as cooperative sensing but without any extra transmission of detection/estimation signal [66]. The principle of HIFI can be briefly explained hereafter. Traditional

decision/detection theory directly maps event to observation and fails to provide the precise event information from the observed physical quantity. This insufficiency becomes obvious when the relationship between the event and the physical quantity is complicated and stochastic. The HIFI decision framework introduces two-mapping concept: from event space to physical quantity (state space), and from state quantity to observation, such that we can model the entire process from sensor observation to decision and subsequent action. To fuse the observations from different physical quantities, the optimal decision is

$$\max_{a} E\left[u\left(a,\theta\right)\left|y_1,\ y_2,\ ..,\ y_K\right.\right] \tag{6.12}$$

While the correlation relationship among physical quantities is generally complicated and unknown, we can execute observation selection and ratio combining to make a decision without the need of precise knowledge of correlation by selecting observation to achieve *maximum a posteriori* expected utility function as

$$\max_{i} \max_{a} E_i\left[u\left(a,\theta\right)\left|y_i\right.\right] \tag{6.13}$$

- Source coding: From the concept of spectrum map, spectrum sensing in a CRN is equivalent to a kind of source encoding of spectrum map. As we described later, source coding without prior knowledge of statistics means universal source coding. Concurrently, spectrum sensing via universal source coding is introduced in [67] with more open in front.

In the most challenging high-density machine swarm, spectrum sensing immediately affects the effectiveness of dynamic spectrum access. Though still in early stage to investigate, efforts start to integrate spectrum sensing with dynamic/opportunistic spectrum access, particularly introducing intelligent methodologies like Markov decision process [68][69], multi-armed bandit [70] and machine learning into the scenario [71 – 73] to iteratively optimize the networking performance. Sensing and information fusion is always a key technology in CONASENSE systems and we provide a theoretical foundation for general sensing by using spectrum sensing as illustration.

6.3.2 Connectivity of Spectrum Sharing Wireless Networks Under Interference

Interference is always a key concern in spectrum sharing networks that is obviously an interference-limited, and even in large ad hoc networks allowing concurrent transmissions for spectral efficiency. Even if the spectrum availability information can be obtained smoothly, we still have an open issue, whether connectivity of spectrum sharing wireless networks is appropriate for networking under interference. We will reveal the answer in the following.

6.3.2.1 Stochastic Geometry Analysis of Inference in a Wireless Network

One of the major challenges for communication in machine swarm or any large wireless network, particularly under the scenario of spectrum sharing, is the characterization of interference. Based on interference analysis, we can ensure concurrent transmissions to maximize the utilization of the given spectrum, under the scenarios of cognitive radio networks or spectrum sharing heterogeneous wireless ad hoc networks. Recent introduction of stochastic geometry analysis supplies a great tool to fulfill this analytical need [74 – 76]. The mathematical model of network interference can therefore be well organized and demonstrated to model cognitive radio networks (CRNs) [74]. Furthermore, the interference in large wireless networks has remarkably explored in [75], to understand some fundamental behaviors in interference-limited wireless networks. As each link in such wireless link may be suffer severe fading and potential outage, random graph model of wireless networks becomes an important analytical tool, while [76] presents an excellent review, based on stochastic geometry, to enable a lot of subsequently useful analysis. There are generally 3 critical factors in such modeling:

i. Spatial distribution of the nodes
ii. Wireless propagation characterization
iii. Overall impact of interferers including mobility and session lifetime

We can always describe network topology via a graph $\mathcal{G} = (V, E)$, defined by the collection of vertices, V, and the collection of edges, E. A vertex represents a node in a wireless network and an edge represents a link between two nodes. Erdös and Rényi pioneer the random graph theory. Among a number of random graph methodologies, we usually model the spatial distribution of the nodes according to a homogeneous *Poisson* point process in the two-dimensional infinite plane. The probability of n nodes being inside a region $\mathcal{R}$ (not necessarily connected) depends only on the total area A_R of the region and is given by

$$\mathbb{P}\{n|\mathcal{R}\} = \frac{(\lambda A_R)^n}{n!} e^{-\lambda A_R}, \ n \geq 0 \tag{6.14}$$

where λ is the spatial density of (interfering) nodes.

Regarding propagation and fading effect, the received power, P_{Rx}, at a distance r from a transmitter with power P_{Tx} is

$$P_{Rx} = \frac{P_{Tx} \prod_k Z_k}{r^{2d}} \tag{6.15}$$

where d is the amplitude loss exponent depending on the environment with typical range from 0.8 (hallways) to 4 (dese urban), and 1 for free-space propagation from point source. $\{Z_k\}$ are independent random variables to account for propagation effects such as fading and shadowing. Earlier model of link availability can also serve the purpose of modeling session lifetime.

The opportunistic nature of spectrum sharing wireless networks (including CRN) and fading statistics of each link in wireless networks suggest random networks for modeling. Details of random network analysis or more precisely random graphical analysis are well documented in recent publications such as [77 – 79]. It is generally believed that Poisson Point Process (PPP) node distribution or Random Geometric Graph (RGG) represents the bottleneck performance of wireless networks.

6.3.2.2 To Hop or Not To Hop

Trials to understand the fundamental limits in wireless (ad hoc) networks have been explored over a decade, e.g. [80]. A breaking through to delineate the capacity of wireless ad hoc networks implicitly suggests the challenge of facilitating large multi-hop networks [81][82]. Following similar approach, the strategy to hop or not to hop in machine swarm has been studied under the scenario of data from machines and sensors to a data aggregator that is typically an access point to wireless infrastructure as Figure 6.1 [83]. In addition to conventional throughput capacity and delay, energy consumption is taken as another consideration. The conclusions consistent of key results in [84] are summarized as follows:

- For a small number of machines (or of high priority traffic), single-hop transmission scheme is suggested as currently 3GPP approach [26]. Machines outside the immediate range of the DAs in the 3GPP LTE-A systems, cooperative access [22] can be adopted, exactly as Section II.
- For a large machine swarm, multi-hop transmission scheme to minimize delay suggests an optimal transmission range $\sqrt{\frac{\log n}{\pi n}}$, where n is the number of machines in the unit circle area.

- For machines requiring low energy consumption and long live time in the machine swarm, the optimal transmission range, $\sqrt{\frac{E_{standby}+E_R}{E_{Tmin}}\left(\frac{\log n}{\pi n}\right)}$, is suggested to minimize the energy consumption, where $E_{standby}$ and E_R denote the standby and receiving power consumption, and E_{Tmin} is the transmission power to reach above optimal transmission range.

Under the scenario of cognitive radio networks or spectrum sharing heterogeneous wireless networks, the precise analysis to provide the answer of "to hop or not to hop" is still unavailable due to complicated interference. However, one recent result investigated the cooperative replay in opportunistic networks [85], to verify the conditions for successful cooperative relay. For dynamic operation of CONASENSE systems, precise working conditions and algorithms are still wanted.

6.3.2.3 Transmission Capacity and Connectivity of CRNs

Since Gupta and Kumar's pioneer analysis [86], the practical limit of wireless ad hoc networks has attracted significant interests. While studying code division multiple access, *outage* well known in wireless communications has been noted as a practical criterion to define the capacity of wireless networks [87]. The consequent *transmission capacity* of wireless networks has been widely accepted as the maximum spatial density of *local* unicast transmissions given an outage constraint multiplying information rate [88], and as a measure of limiting performance of a wireless ad hoc network. In other words, for a maximal outage ε and information rate R, the maximum spatial density of transmissions is obtained as λ, and then the transmission capacity is $\lambda(1-\varepsilon)R$. With general fading, transmission capacity of wireless ad hoc networks is difficult to obtain. However, upper bound was developed in [89] and later upper bound of mean node degree and lower bound on node isolation probability were also analytically derived under general fading [90]. Although the transmission capacity for CRN or spectrum sharing heterogeneous wireless ad hoc networks is still in need so that we can ensure the meaning of communications in machine swarm, we still can somewhat realize such feasibility. The more critical issue to facilitate spectrum sharing ad hoc networks might be analytical understanding of connectivity, the degree for each node in the network connecting to neighboring nodes. Obviously, the connectivity is so critical to design routing algorithms and many other networking functions, and is so difficulty to analytically tackle this challenge. Further intellectual breaking through is surely required.

In addition to stochastic geometry, the relationship of random graphs and statistical mechanics has been amazingly noted [84][92], which plays a key role in subsequent study using the concept of *percolation* in statistical mechanics [93]. The intuitive meaning of percolation in statistical mechanics explains quick phase transition, say from liquid water to solid ice in a quick, large-scale and homogeneous way once temperature dropping to 0℃. In wireless networks, we can use Figure 6.5 to intuitively explain percolation in wireless networks. Once reaching the percolation threshold, the entire wireless network can quickly get connected away from original rather isolated configuration. Please note the change of transmission range to result in almost full connection in the figure that is similar to what we see in condense matters as a branch of physics. The efforts trying to analyze connectivity in multi-hop wireless networks are actually not new [94][95], however, it is still distant for spectrum sharing cooperative ad hoc networks (or at least CRN), which is important in facilitation of communications in machine swarm.

Based on stochastic geometry, a pair of nodes gets connected if the channel capacity of the link between them is greater or equal to target information rate. Through the interference analysis to get percolation threshold, we can obtain the degree of node in semi-closed form for CRN and thus optimal power allocation to maximize the effective mean degree of a secondary user node in CRN [96]. In the same paper, it is also extended to obtain the results

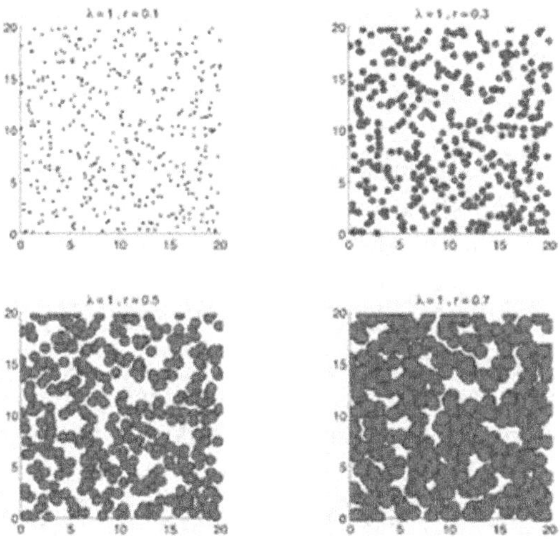

Figure 6.5 When λ Denotes Node Density and r Denotes Transmission Range, Percolation (from Rather Isolated Connections to Pretty Full Connections) Happens. (Taken from [91])

for multiple cooperative spectrum sharing heterogeneous ad hoc networks as the most general case. By looking some typical parameter settings, we may conclude the positive feasibility of spectrum sharing multi-hop networking for machine swarm. Based on such connectivity analysis, we can further show that a node can realize cooperation from hybrid and interconnected wireless networks and wireless infrastructure [97]. It suggests that spectrum sharing multi-hop networking in machine swarm, as Figure 6.1, is definitely possible as a firmed answer for major challenges at beginning of this section. In front of us, there are a great number of new challenges of exact ways to design such networking in machine swarm.

6.3.3 Routing in Cooperative Cognitive Ad Hoc Networking

To implement cooperative multi-hop (ad hoc) networking for machine swarm, routing might be the first challenge into system design. Routing for ad hoc networks has been widely studied for a long time [44], but generally requires end-to-end routing information to establish the routes, which is not feasible in spectrum sharing machine swarm at all, due to spectral efficiency and likely unidirectional opportunistic availability [39]. We use Figure 6.5 to illustrate multi-hop cognitive radio networking in machine swarm or sensor networks. In this scenario, a source CR (denoted as node n_S), a destination CR (denoted as n_D), and several relay CRs (denoted as n_Rs) that can cooperatively relay packet flowing from n_S to n_D. We assume there are n relay CRs. In order to avoid the interference to PSs, CSN links are available under idle duration of PSs that DSA can effectively fetch such opportunities, after successful spectrum sensing. Link available period in CSN results in random network topology even all nodes being static. We assume that the n_S, the n_D, and the n_Rs can be mobile. Suppose that there are K possible opportunistic paths between n_S and n_D. The set of total K opportunistic paths is denoted as $\mathcal{P} = \{p_1, p_2, \ldots, p_K\}$, where the ith opportunistic path p_i consists of J_i links, for $i = 1, 2, \ldots, K$. n_S transmits a set of data $X_1, X_2, \ldots, X_K$ over these K paths. The values of the data are observed from some joint distribution and can be either continuous or discrete.

Cooperative routing has been introduced into sensor or energy-sensitive ad hoc networks, though based on Gaussian-Markov field [98] or static grid networks [99]. With the help of spectrum map (no need to be perfectly known), the spectrum aware routing is first proposed consisting of two parts: leveraging opportunistic routing based on spectrum map as local routing (among neighborhood), and global routing following the trend of spectrum map via greedy routing [100]. Spectrum map can actually indicate strong interference area with the help of tomography even an incomplete spectrum map [101]. If such spectrum map is constructed by compressive sensing, it is shown to achieve successful routing by considering the information needed to route [64]. Routing under insufficient spectrum map information is critical to routing in machine swarm as impossible to obtain complete spectrum information for each machine under spectrum efficiency. The value of [64] lies in demonstration a simple routing algorithm proceeding on incomplete information, but reach the goal *statistically* speaking. Such a statistical concept could be extremely useful to implement machine swarm communications, as we can not execute precise control but we can achieve our goal in a statistical manner in a large network. In parallel, applying interference information to form a routing game is studied in [102], and local routing in dense RGG sensor networks is investigated in [103].

Similar to diversity in wireless communications, *route diversity* can greatly improve reliability of ad hoc networks [104] and surely spectrum sharing multi-hop networks, which is also known as *multi-path routing* that is precisely depicted in Figure 6.6. *Network coding* is a branch of information theory to optimize information flow in a network [105]. We can therefore apply simple network coding to greatly improve the rate-delay operation in ad hoc networks [106], and further reliability with rate-delay performance [107]. Such multi-path networking will be our foundation of statistical control and error control over cognitive radio ad hoc networks in next sub-section. As optimization can

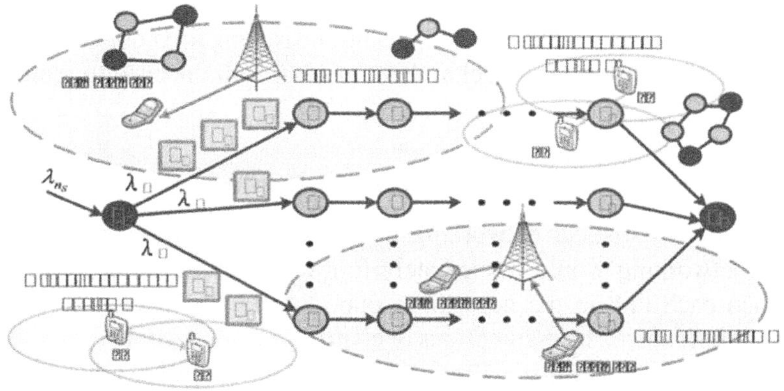

Figure 6.6 Multi-hop Cognitive Radio Multi-path Networking in Machine Swarm

be conducted for delay or for energy, it is reasonable to conjecture that energy aware routing shall equivalently proceed [99][108].

Under the thinking of network operating under incomplete information, *statistical* control of networks has been developed as the way that must go and will be described in detail. However, *statistical routing* is still a missing part in spite of above theoretical ground. Under a slightly different motivation to consider the randomness of channels and interference, statistical routing is developed in [109]. Assuming perfect spectrum sensing to know the SINR of each transmitter-receiver pair, we again consider data *percolation* through a wireless network by the packet delivery *probabilities* to optimize routes, transmission probabilities, and corresponding transmission power. For totally N nodes to transmit, the optimization becomes

$$\max \sum_{n=1}^{N} U_n\left(r_n\right) - \sum_{n=1}^{N} C_n\left(\rho_n\right) \tag{6.16}$$

subject to various physical constraints, where the utility function $U_n\left(r_n\right)$ is a function of rate r_n and cost function $C_n\left(\rho_n\right)$ is a function of physical layer transmission parameters. A successive convex approximation is beautifully developed to find a Karush-Kuhn-Tucker (KKT) solution for the optimization to complete statistical routing. How to combine above together would be a great technical challenge to facilitate statistical routing in machine swarm, which have to proceed based on local information (i.e. no end-to-end information) only but achieve the purpose of communications in terms of statistical performance. Please also note further challenges for CONASENSE scenarios of mobility, like vehicular and robotics. Sensing and routing for mobile devices is a hardly touched research subject at this time, but badly needed to realize applications.

6.3.4 Statistical Control of QoS and Error Control

After introducing statistical networking in previous sub-section, *statistical control* of networking would immediately follow as next technology challenge. When each link in the network is opportunistically available under fading and interference, the assumptions of control channel, feedback, end-to-end information, are not realistic. To avoid tremendous control overhead, the rationale is to statistically control the target performance in a large network without micro-management to reach satisfactory performance in statistics, similar to the law of large number or large deviation [110]. In such a way of statistical control, we may consume more resource to create diversity at the beginning but may actually save resource finally and more reliable [104] due to the saving of control overhead.

We first consider the methodology to retain quality of service (QoS) in CRN. Please recall Figure 6.5, if we allow source node to proceed multi-path forwarding to against random link availability due to (i) opportunistic spectrum access (ii) outage by fading and interference (iii) packet violation of time-to-live, we may apply the effective bandwidth discussed in section 2 [33 – 36] to statistically retain delay control as [111], and further statistical QoS provisioning in interference-limited underlay CRN [112]. It is also possible for heterogeneous services in CRN [113]. We may imagine an end-to-end "session" on top of these multi-paths. This achievement is particularly useful to applications like video surveillance and wireless robotics.

Although end-to-end feedback sounds impossible in machine swarm of cognitive radio capability, the end-to-end error control is still desirable. Again leveraging multi-path forwarding, we may implement hybrid ARQ (HARQ) into packets to different paths. Although there might not be able to successfully receive desirable packets from all paths, this is exactly as the operation of HARQ and thus achieves end-to-end error control in CRN [114]. Similarly, we may apply error control for local communication in heterogeneous ad hoc networks, to improve the connectivity and thus networking performance in the ad hoc network [115].

A more exciting scenario can be viewed by just looking at n_S and n_D, while the multi-path multi-hop networking is treated as an aggregated "multi-input-multi-output channel" (3×3 in Figure 6.5). n_S and n_D equivalently have multiple antenna to/from this aggregated "channel", to form a *virtual MIMO* on top of this imaginary "session", different from traditional definition of virtual MIMO controlled by network. Via such abstract mapping, we can modify the well-known space-time codes into *path-time codes* (PTC) [116]. By creating appropriate encoding matrix (permutation coding matrix or maximum-distance coding matrix), we modify sphere decoding to achieve significant error correcting gain. PTC can be generally applied to CRN or any ad hoc networks, which shall enable a new design paradigm in ad hoc networks, reliable end-to-end performance without control overhead and feedback in each link, a significant advantage for spectrum and energy efficiency.

More control issues in CRN can be found in literature such as queue control for service interruption [117], topology control [118], congestion control [119], and call admission control [120]. To practically implement spectrum-sharing communication in machine swarm, tremendous technical opportunities are required for various control mechanisms.

6.3.5 Heterogeneous Network Architecture

Although we demonstrate the feasibility of spectrum sharing multi-hop networking in section 3.2, we still have to make sure whether such an approach is effective, particularly a packet of reasonable delay from source to destination. In large networks, this is not easy as we can abstractly consider from the social network analysis or random graph analysis. The *network diameter* is a good indicator of the required number of hops between two nodes in a network [SN]. Although network diameter has been widely studied [77], PPP network arrangement of high interests in wireless networks was not known. Unfortunately, in machine swarm, network diameter analytically derived from PPP network topology suggests intolerable number of hops for communications in machine swarm [121].

To resolve this challenge of autonomous M2M communications, we may ironically recall a result from social network research, to reflect a novel truth of interplay between technological networks and social networks [122]. A famous theory in social networks is the 6-degree separation for any two persons in the world, a large human "network". Obviously, the network diameter of this large human network cannot be 6. By digging into this *small world phenomenon*, a *short cut* always exists to link two far separated persons within the neighborhood [123]. Suggested in [121], we shall establish a man-made short cut among data aggregators (DAs) in Figure 6.1. Luckily, such a short cut or like an information expressway is actually available as part of wireless infrastructure and cloud networking given in the heterogeneous network architecture of Figure 6.1 and discussed in Section II. The study of message delivery time is explored in [121] to show significant performance improvement, and general study in [124]. The rest of challenge is placement of DAs studied in [121] to show satisfactory performance adopting uniform distribution. Extending earlier QoS control techniques, we enjoy statistical QoS guarantees in machine swarm without detailed microscopic and end-to-end control [125].

Heterogeneous network architecture involves a lot of system design issues under active research at this time for both 3GPP cellular networks and IP-based wireless networks such as wireless LANs. It also suggests new opportunity to develop software defined network architecture to allow scalable deployment.

6.3.6 (Information Dynamics and) Traffic Reduction and In-Network Computation

Data gathering/collection [126], distributed detection and data fusion [127], data analysis [128], and data aggregation [129], have been extensively studied for wireless sensor networks of different purposes but primarily for the efficiency of processing data. In light of large wireless networks, a novel technology known as *in-network computation* was proposed to process data before reaching final destination in large wireless sensor networks [130]. It suggests a useful concept to combine computation and communication in a sensor network, to form a new design paradigm as computing compromising computation and communication (i.e. moving data). Further applications of in-network computation in wireless sensor networks can be found in [131], as a sort of practical realization of *context computing*. However, to deal with the fundamental spectrum efficiency in large networking for machine swarm is still missing in technology development.

Along the development of network coding and peer-to-peer networking, an interesting idea was developed by considering a special and ideal case to identify the optimal way to send n packets among n transmitter-receiver pairs using the nature of broadcasting in wireless networks [132]. The goal is to minimize the total traffic load or total energy, however, please note that minimizing traffic suggests another way to enhance spectral efficiency if we consider a practical definition of spectral efficiency as *network throughput per bandwidth*, which serves a side benefit from original research. The facilitation of this special case is via source coding into networking scenario, *network-aware source coding* or more precisely Hoffman coding. This nice idea is subject to further generalization, as the unrealistic assumption in Hoffman coding is *prior* knowledge of probability distributions that means each node knowing the entire network information. Universal source coding [133] serves the purpose without the need of priori statistics. Focusing on data gathering in wireless sensor networks, joint opportunistic source coding and opportunistic routing has been remarkably developed [134]. Pointed in sub-sections 3.3 and 3.4, network coding enhances performance by mixing information contents prior to forwarding packets. This research further explores joint source coding and (opportunistic) routing in a successful way via leveraging correlation of data in data gathering. Intuitively, the performance gain comes from (i) multiple independent receiving equivalent to diversity gain (ii) compression by Lempel-Ziv universal source coding when two correlated packets are received by one node. Unfortunately, this scheme assumes a node knowing the average number of hops for all forwarding candidates. In a large ad hoc network, efforts are still required for practical applications. Another experimental study is reported by considering finite alphabets for the data to be collected and transmitted in multi-hop networking [135]. By leveraging the broadcasting nature and information fusion like compression, it is shown that certain portion of relays are not needed anymore to significantly reduce the traffic in a network to equivalently increase the effective spectral efficiency (i.e. allowing more packets to transmit for other purposes). For large machine swarm, this provides an alternative to significantly enhance spectral efficiency (i.e. network throughput per bandwidth), though control information is still required. Under autonomous CONASENSE operating, effective information-centric processing to enable energy and spectrum efficiency would be a wide-open technology opportunity by confining overhead of control signaling and mechanism.

The control information in a large network can consume tremendous bandwidth and opposite to spectral efficiency. Figure 6.7 depicts the possibility of traffic reduction in information collection multi-hop sensor networks.

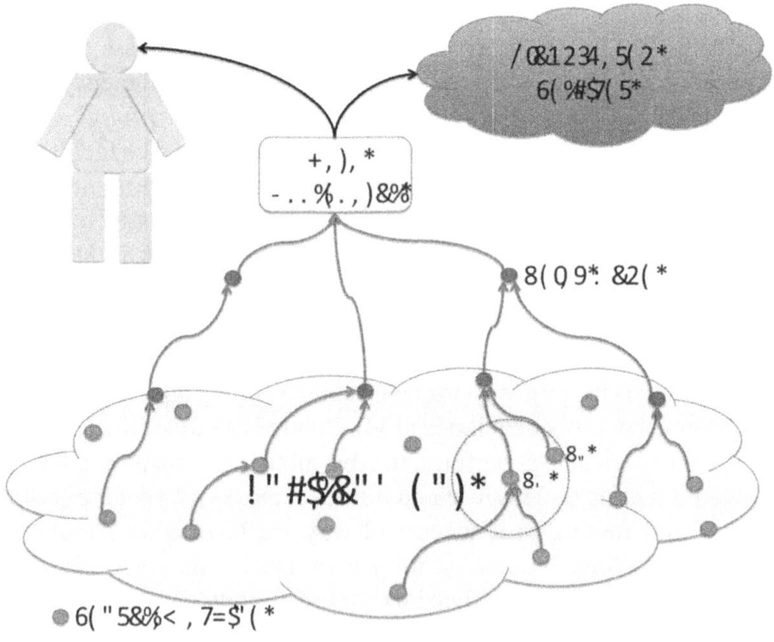

Figure 6.7 Traffic Reduction Leveraging the Nature of Broadcasting

Before machine R_m to transmit, it receives messages from two other machines and then compresses into a new message together with its own information collection. For machine R_n collecting information from the field, because R_n is within the transmission range of R_m indicated by red circle, it can compute the extra information to transmit and possibly realizes no extra information to transmit based on what R_n heard from R_m, which can this possibly save transmission bandwidth. Instead of further information theoretical study, traffic reduction (or equivalent transmission energy reduction) in the data collection wireless networks can be formulated as straightforward fusion together with quantization (i.e. compression) [136]. Then, we can show that a node can overhear from other nodes due to the nature of broadcasting to execute compression and thus to effectively save (or eliminate) portion of relay transmissions in a network while keeping the same precision of estimation. However, instead of the assumption on control information of entire network in this approach, this approach assumes that the multi-hop tree network topology that still requires efforts to establish in advance though each node does not require *prior* knowledge of network topology. Consequently, traffic reduction in general multi-hop networks remains an open problem.

However, in network computation and source coding are not only useful in traffic reduction for better spectral efficiency (or, equivalently energy efficiency for less relays). A practical and therefore desirable design in large machine swarm is to reduce end-to-end delay and/or energy efficient networking, without the required knowledge of end-to-end routing information at each node nor any prior knowledge of network topology. As shown in Figure 6.5, n_S transmits a set of data $X_1, X_2, \ldots, X_K$ over these K paths. The values of the data can be observed from some joint distribution. Such observations may not consume bandwidth as a part of spectrum sensing or CRN tomography [60]. In-network computation for distributed source encoding and linear block coding, together with greedy routing can actually greatly enhance spectrum utilization or minimize end-to-end delay in machine swarm of cognitive radio capability [137], to indirectly enhance spectral efficiency. In new trend of applying source coding into routing, we therefore introduce a new way to execute in-network computation for the benefit of communication resource by leveraging the broadcasting nature of wireless communications without the need of global networking information.

6.3.7 Nature-Inspired Approaches toward Time Dynamics of Networks

For decades since the birth of computer networks, network functions are designed typically based on a snapshot at certain timing and then to develop algorithms to optimize performance. The benchmark methodology via optimization is summarized in [138] as the state-of-the-art network design. However, we are more interested in time dynamics of networks, particularly multi-hop networks, as networks are not expected to operate in steady state. Before exploring time dynamics of spectrum sharing networks, please recall another new design paradigm from several places in Section 2 of this chapter: to treat *radio resource* as the core of cross-layer design, such as the heterogeneous network design for cyber-physical system based on allocation of radio resource blocks for either access or interference mitigation.

Following the cross-layer design philosophy centered by radio resource allocation, the spectrum sharing wireless networks, no matter CRN or heterogeneous ad hoc networks, can be view as nodes/users to share radio resource, which is pretty much similar to predator-prey relationship. Radio resource units correspond to preys and network users/nodes are just like predators to consume radio resource (i.e. preys). Prey-predator population dynamics are well studied by mathematicians in past centuries. Appropriate

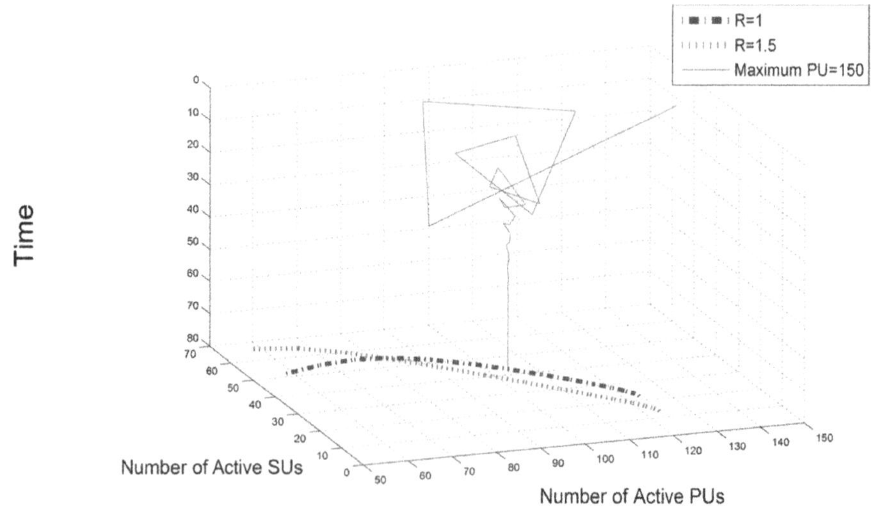

Figure 6.8 Phase Diagram for Time Dynamics of CRs Sharing Spectrum [139]

modifications can perfectly represent CR systems [139]. Most importantly, the time dynamics of CRs can perfectly described such as Figure 6.8 showing system evolution with time and steady state performance (also suggested spectrum utilization). This amazing approach is just a starting point. Further development by including access protocols and graphical analysis is expected to reveal more fundamental behaviors of spectrum sharing machine swarm networking and common wireless networks.

6.4 Energy-Efficient Implementation, Security and Privacy, Network Economy, Deployment and Operation

Following above explorations on the fundamental technology of M2M communications, more practical aspects have to be considered toward realistic implementation, deployment, and operation.

6.4.1 Application Scenarios of M2M System

M2M along with Internet of Things extends human life in a diverse way. Facilitation of autonomous M2M communications to handle the interaction of cyber and physical systems, in addition to fundamental technologies, relies on corresponding application scenarios [140–145]. Namely but not an exhaustive list,

- Smart home/office
- Smart community and smart city
- Environmental and ecology monitoring for safety, disasters, agriculture, etc.
- Surveillance
- Energy-efficient control such as smart grid
- Healthcare
- Factory automation
- Intelligent vehicles
- Wireless robotics

Such applications suggest further technology opportunities of autonomous M2M communications in hardware, software, and system integration.

- Energy efficient wireless communications and sensor networks
- Spectrum efficient communications and networks
- Scalable communication networks

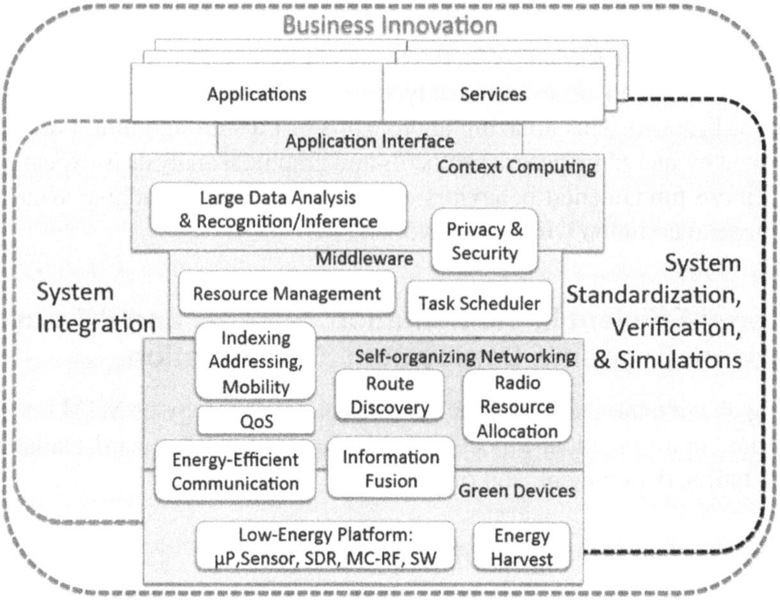

Figure 6.9 Functional Block Structure of M2M Communications for CONASENSE

- Information-centric networks (with capability of information fusion, in-network computation, and data analysis/mining)
- Addressing and index, navigation, mobility management, service discovery, and middleware
- Security and privacy, particularly in heterogeneous networking environments
- Green devices for sensing, communication, and computing

We use the following block diagram as Figure 6.9 to illustrate potential M2M system function, while the role related to M2M communication and interaction with entire CONASENSE system has been also delineated. In addition to those technologies that we described in earlier sections as the framework of M2M communications, there are some more of our particular interests in later of this section.

6.4.2 Energy Harvesting Communication Networks

An emerging technology is the design of new communication systems and networks due to machines and sensors operating on harvested energy. For tremendous amount of machine devices in M2M systems, battery operating

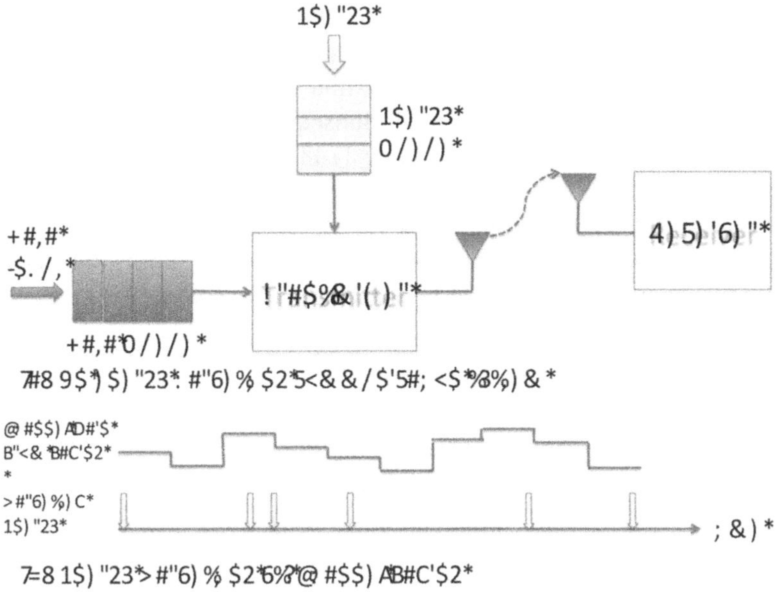

Figure 6.10 Stochastic Model for An Energy-Harvesting Communication System (a) communication system (b) energy harvesting vs. fading

hardware can create a great challenge for device management and environmental hazard. As a matter of fact, a device can realistically harvest energy from natural environment through sunlight, vibration, electromagnetic radiation, wind, etc. to operate [38]. Such energy source usually does not support streamline and stable power-supply and usually implies limited storage time and amount of energy. Any communication link relying on energy harvest is therefore opportunistic somewhat like cognitive radio, while the opportunistic nature of cognitive radio comes from spectrum availability and that of energy harvest communications comes from energy availability. Currently, such opportunistic nature is commonly modeled as a random process to give a general stochastic modeling for an energy harvesting communication systems as Figure 6.9, and consequently queuing model for energy harvesting networking.

Since the energy harvesting can be viewed as a random process along the time axis, just like a fading channel, the immediate optimization of transmission targets on transmission power [146], and such optimization is equivalent to famous water filling in information theory [147]. A special feature in energy harvesting communication is the constraint on time resulted from energy storage on a machine. Transmission completion time shall be minimized, from the point of view not only from physical-layer communication link but also data networking. Given a deadline T, the maximum departure $D(T)$ is defined given energy arrivals and fading channel. Subsequent optimization methods are obtained for link [147] and scheduling [148]. Further development on broadcasting for multiple transmitter and multiple receivers [149] and optimal transmission for re-chargeable energy harvesting battery [150] are achieved.

In the early stage of technological exploration, some immediate open issues of energy harvesting M2M communications may include:

- Statistical model(s) of energy harvesting
- Cross-layer communication system design
- Optimization of communication and computation energy in a machine
- Networking protocols and algorithms optimizing energy and spectrum utilization

In addition to short-term optimization in energy harvesting communications, further machine learning to enhance long-term operation is also considered [151]. Opportunistic communication networking is still open at this moment.

6.4.3 Security and Privacy

Cyber security and privacy has been attracting great amount of technical interests in recent years, which we do not want to repeat here. However, machine-to-machine communications involve interactions between cyber and physical worlds, which indeed introduces a lot of new issues in security and privacy [152]. In this paper, among this wide subject area, we orient two newly merging challenges in security and privacy related to M2M and CONASENSE communications.

Due to the co-existence of two kinds of networks, cyber and physical, *inference attacks* can play a much more powerful role in privacy and security. Location privacy in mobile communications was initially noted [153]. Later, further inference attacks are suggested. For example, by investigating household electricity consumption along the time through listening a smart meter in smart grid M2M communications, a lot of private life patterns can be inferred logically or statistically. Such a privacy attack and thus privacy preservation methodology can be understood and modeled by utility function and source coding [154] in a novel way.

Another sort of attacks is the distributed denial-of-service (DDoS) attack, which has been first introduced to cellular communications in [155] and can be harmful by easy jamming autonomous M2M communications to disable the operation of physical systems. Such security treat can be magnified by more factors when spectrum sharing M2M communications are conducted, primarily: (i) the existence of phase transition in heterogeneous CRN suggests sensible operating point change in network operation [91] (ii) a spectrum sharing network can be viewed as an eco-system and each node/machine tends to be selfish to evolve its own operating strategy [156]. Therefore, in addition to known attacks, more powerful and hard-to-detect DDoS attacks can be implemented [157]. The defense strategy can leverage information fusion in the network and form a game to retain network resilience [157][158].

6.4.4 Spectrum Sharing Network Economy

As pointed in Section 3 regarding *layerless* "cross-layer" system design for M2M communications in the machine swarm, radio resource allocation (i.e. spectrum management) becomes the key, which is nice in technology but we have to examine practical feasibility in business operation to fulfill the economic need of M2M system operator. In other words, whether practical operation of spectrum sharing (or cognitive) machine swarm communication is realizable depends on network economy. This concern starts from technological network design by investigating resource allocation [159] by confining the problem as the convex optimization of radio resource in wireless cognitive networks [160]. Most recently, such efforts extend to network coding assisted channel allocation for routing [161]. With more practical economy study into scenario later, the spectrum sharing network economy actually must satisfy four parties: primary system users, secondary cognitive radio users, operator of the primary system and secondary system, and regulator who cares overall spectrum utilization and users' QoS. An auction mechanism is established to ensure such satisfaction of network economy [162]. Such research can be extended to heterogeneous spectrum sharing wireless networks [163] and even cellular systems [164], which suggests economic feasibility of multiple M2M communication operators sharing certain spectrum. Finally, spectrum sharing mechanism may happen between different industries, broadcasting and wireless communications as Figure 6.11 demonstrating.

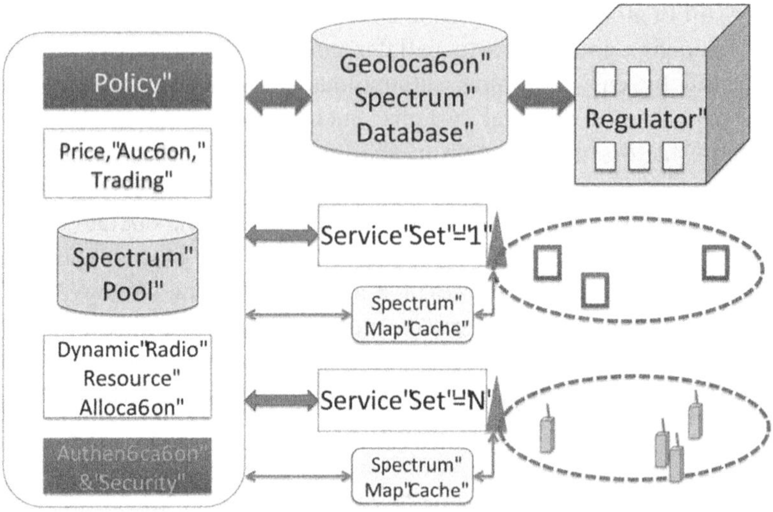

Figure 6.11 Spectrum Sharing Between Digital Broadcasting and Wireless Communications

6.4.5 Implementation, Deployment, and Sustainable Operation

In spite of a lot of explorations on realization of efficient M2M communications, particular challenges in machine swarm. However, we have to look at some practical issues to install, to deploy, and to operate in a sustainable way. For wireless infrastructure, it is an extension of existing well operating cellular-type systems, and there shall not be any major fundamental technological obstacle in spite of tremendous technical efforts in need. Backward compatibility and scalability for potential expansion to accommodate DAs to support large number of machines might be something easy to overlook. Using smart coding, such an access mechanism to DA can be retained as [165].

There are further implementation concerns regarding spectrum sharing cooperative multi-hip networking in machine swarm:

- Time synchronization: It is too spectrum-costly to maintain a control channel in machine swarm communication. Typical thinking is to assume time synchronization in CRN to avoid performance loss. Ways to establish network synchronization can adopt algorithms from consensus and cooperation [166] like gossip algorithm etc. In large network, the synchronization overhead can be significant and further stabilization is not fully understood. However, a different thinking to dynamic access the radio resource in asynchronous way is suggested [167]. Based on game theoretical formulation for each machine, even better performance than synchronous access can be surprisingly achieved. Asynchronous spectrum access can open a brand new door for advances in spectrum sharing network technology.

- Complexity to implement cognitive radio and spectrum sharing networks: The complexity of CR implementation is widely considered high. However, a good example is the adaptive frequency hopping (AFH) [168] in IEEE 802.15.2 and Bluetooth 2.0 and later versions. AFH requires to sense the spectrum and then to decide the change of hopping sequence to avoid collisions, such that synchronous (SCO) stream-traffic link can be well maintained among different networks to share the same spectrum. This is pretty much aligned with the principle of cognitive radio or spectrum sharing scenario in machine swarm. However, as long as we can select parameters in an appropriate manner and design accordingly, the incremental complexity in implementation is actually very minimal that can be ignored. With this energy-efficient technology has been widely used for nearly a decade with billions of users, CRN is really just a technology challenge in effective system design.
- Processor architecture for devices: A real fundamental technology challenge lies in energy-efficient processor to effectively support data handling and M2M communication/networking while fitting simple machine or sensor. Among many literatures looking into this subject, we would like to remind a different design paradigm. By looking into communication algorithms, we may note only a few fundamental computation modes governing successful executions, and a scalable processor can be design to reach balance of performance and power consumption [169]. With the need of no more than 20 instructions to execute software, such programmable processor of programming elements for machines/devices in the swarm implies a coming design paradigm shift for both hardware and software.

For successful operation of machine swarm, we have to note the device management caused by unreliable operation lifespan [143][170] and massive amount of data under cloud-based systems [171]. Finally, a remaining technological challenge in this paper is that reported technical achievements related to communication in machine swarm are most based on cognitive radio networks, and have not been generalized to heterogeneous spectrum sharing ad

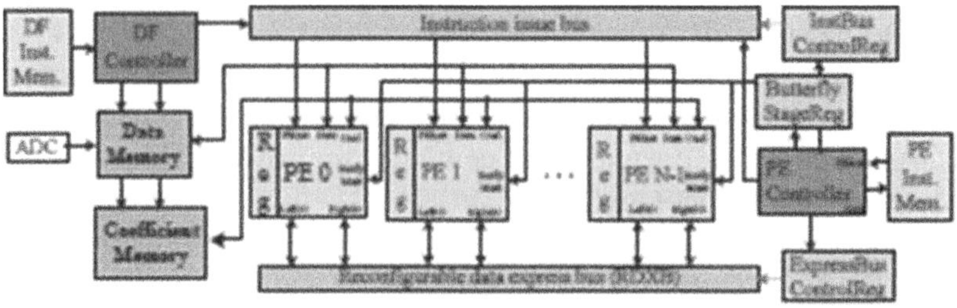

Figure 6.12 Green Communication Processor of Scalable Programming Elements

hoc networks yet. It is still an open research territory of human's engineering knowledge.

6.4.6 Toward the Reference Model of M2M Communication Architecture

The purpose of reference model is to enable easy collaboration during practical implementation and operation among sub-systems, and between hardware and software. M2M communications supports autonomous information (including data and control) transportation within the entire cloud-based IoT systems or cyber-physical systems. Therefore, M2M communication reference model shall allow transparency to application software and scalable to hardware expansion. In spite of tremendous application scenarios in Section 4.1, [172] suggests that the fundamental communication/networking styles are rather limited and we may summarize as follows:

- Streamline traffic of variable rate or fixed rate
- Periodic traffic
- Bursty traffic
- Arbitrary traffic

Communication patterns implies different functions of M2M communication reference model, while keeping in mind that we wish to minimize the amount of control signal due to tremendous number of machines. Such reference model transparent to applications shall play the central role in hardware and software system integration, linking cyber world and physical systems [174] by referencing Figure 6.9. In particular, ad hoc networking in machine swarm based on local information (i.e. no global information) to achieve performance in a statistical manner is still subject to further research on precise system design. We are expecting to international standardization on top of an appropriate as M2M communication reference model as the final open technological challenge to facilitate common hardware platform of sensing, communication, fusion and decision on the collected data. Subsequently, universally applicable middleware and scalable application software could be developed.

6.5 Concluding Remarks

This chapter presents state-of-the-art technologies for the entire M2M communications and remaining intellectual and engineering challenges. With effective sensing and consequent communication capability, we do foresee tremendous potential controlling M2M plays a central role to benefit modern and future human life, to reach the ultimate goal of CONASENSE.

Acknowledgement

The authors would appreciate Dr. J. Chris Ramming and Dr. Shu-Ping Yeh, INTEL Research Labs. for their constructive suggestions. This research was supported in part by the National Science Council, National Taiwan University, and Intel Corporation under the grants NSC101-2911-I-002-001 and NTU102R7501, and IBM Shared University Research Award 2012–2013.

References

[1] E.A. Lee, "Cyber Physical Systems: Design Challenges", 11th IEEE Symposium on Object Oriented Real-Time Distributed Computing, 2008.
[2] R. Rajkumar, I. Lee, L. Sha, J. Stankovic, "Cyber-Physical Systems: The Next Computing Revolution", ACM Design Automation Conference, 2010.
[3] L. Atzori, A. Iera, G. Morabito, "The Internet of Things: A Survey", Computer Networks, vol. 54, pp. 2787–2805, 2010.
[4] G. Wu, et al., "M2M: from Mobile to Embedded Internet", IEEE Communications Magazine, vol. 49, no. 4, pp. 36–43, April 2011.
[5] S.Y. Lien, K.C. Chen, Y.H. Lin, "Toward Ubiquitous Massive Accesses in 3GPP Machine-to-Machine Communications", IEEE Communications Magazine, vol. 49, no. 4, pp. 66–74, April 2011.

[6] T. Taleb, A. Kunz, "Machine Type Communications in 3GPP Networks: Potential, Challenges, and Solutions", IEEE Communications Magazine, vol. 50, pp. 178–184, March 2012.
[7] K. Zheng, W. Xiang, M. Dohler, "Radio resource Allocation in LTE-Advanced Cellular Networks with M2M Communications", IEEE Communications Magazine, vol. 50, no. 7, pp. 184–192, July 2012.
[8] 3GPP TS 22.368 V12.0.0, "Service Requirements for Machine-Type Communications," Sept. 2012.
[9] 3GPP TR 23.888 V11.0.0, "System Improvement for Machine-Type Communications," Sept. 2012.
[10] Damnjanovic, J. Montojo, Y. Wei, T. Ji, T. Luo, M. Vajapeyam, T. Yoo, O. Song, and D. Malladi, "A Survey on 3GPP Heterogeneous Networks," IEEE Wireless Commun., vol. 18, No. 3, pp. 10–21, Jun. 2011.
[11] 3GPP TS 36.300 V11.3.0, "Evolved Universal Terrestrial Radio Access (E-UTRA) and Evolved Universal Terrestrial Radio Access Network (E-UTRAN); Overall description," Sept. 2012.
[12] D. López-Pérez, I. Güvenç, G. D.L. Roche, M. Kountouris, T. Q.S. Quek, and J. Zhang, "Enhanced Intercel Interference Coordination Challenges in Heterogeneous Networks," IEEE Wireless Commun., vol. 18, No. 3, pp. 22–30, Jun. 2011.
[13] S.Y. Lien, S.Y. Shih, and K.C. Chen, "Spectrum Map Empowered Resource Management for QoS Guarantees in Multitier Cellular Networks," in Proc. IEEE CLOBECOM, 2011.
[14] S.Y. Lien, Y.Y. Lin and K.C. Chen, "Cognitive and Game-Theoretical Radio Resource Management for Autonomous Femtocells with QoS Guarantees," IEEE Trans. Wireless Commun., vol.10, no.7, pp.2196–2206, Jul. 2011.
[15] X. Y. Wang, P.H. Ho, and K.C. Chen, "Interference Analysis and Mitigation for Cognitive-Empowered Femtocells Through Stochastic Dual Control," IEEE Trans. Wireless Commun., vol.11, no.6, pp.2065–2075, Jun. 2012.
[16] S.M. Cheng, S.Y. Lien, F.S. Chu and K.C. Chen, "On exploiting cognitive radio to mitigate interference in macro/femto heterogeneous networks," IEEE Wireless Commun., vol.18, no.3, pp.40–47, Jun. 2011.
[17] Y.S. Liang, W.H. Chung, G.K. Ni, I.Y. Chen, H. Zhang, and S.Y. Kuo, "Resource Allocation with Interference Avoidance in OFDMA Femtocell Networks," IEEE Trans. Veh. Technol., vol.61, no.5, pp. 2243 - 2255, Jun. 2012.

[18] Y. Sun, R.P. Jover, and X. Wang, " Uplink Interference Mitigation for OFDMA Femtocell Networks," IEEE Trans. Wireless Commun., vol.11, no.2, pp. 614 - 625, Feb. 2012.

[19] Ö. Bulakci, S. Redana, B. Raaf, and J, Hämäläinen, "Impact of Power Control Optimization on the System Performance of Relay Based LTE-Advanced Heterogeneous Networks," Journal of Commun. & Networking, vol.13, no.4, pp. 345 - 359, Aug. 2012.

[20] R. Combes, Z. Altman, and E. Altman, " Self-Organizing Relays: Dimensioning, Self-Optimization, and Learning," to appear in IEEE Trans. Network & Service Management, 2012.

[21] Z.M. Fadlullah, M.M Fouda, N. Kato, A. Takeuchi, N. Iwasaki, and Y. Nozaki, "Toward Intelligent Machine-to-machine Communications in Smart Grid,", IEEE Commun. Mag., vol.49, no.4, pp.60–65, Apr. 2011.

[22] S.Y. Lien, T.H. Liao, C.Y. Kao and K.C. Chen, "Cooperative Access Class Barring for Machine-to-Machine Communication," IEEE Trans. Wireless Commun., vol.11, no.1, pp.27–32, Jan. 2012.

[23] K.S. Ko, M. J. Kim, K.Y. Bae, D.K. Sung, J.H. Kim, and J.Y. Ahn, " A Novel Random Access for Fixed-Location Machine-to-Machine Communications in OFDMA Based Systems," IEEE Commun. Letters, vol.16, no.9, pp.1428–1430, Sept. 2012.

[24] ZTE, "R2–104662: MTC Simulation Results with Specific Solutions," 3GPP TSG RAN WG2 Meeting 71, Aug. 2010.

[25] CATT, "R2-100182: Access Control of MTC Devices," 3GPP TSG RAN WG2 Meeting 68bis, Jan. 2010.

[26] S.Y. Lien, K.C. Chen, "Massive Access Management for QoS Guarantees in 3GPP Machine-to-Machine Communications," IEEE Commun. Letters, vol.15, no.3, pp.311–313, Mar. 2011.

[27] S. E. Wei, H. Y. Hsieh, and H. J. Su, "Enabling Dense Machine-to-Machine Communications through Interference-Controlled Clustering," in Proc. IEEE IWCMC, 2012.

[28] C.Y. Ho and C. Y. Huang, "Energy Saving Massive Access Control and Resource Allocation Schemes for M2M Communications in OFDMA Cellular Networks," IEEE Commun. Letters, vol.1, no.3, pp.209–211, Jun. 2012.

[29] 3GPP RAN WS, "RWS-120003: LTE Release 12 and Beyond," in 3GPP RAN WS on Rel-12 and onwards, Jun. 2012.

[30] K. Doppler, M. Rinne, C. Wijting, C.B. Ribeiro, and K. Hugl, " Device-to-Device Communication as an Underlay to LTE-Advanced Networks," IEEE Commun. Mag., vol.47, no.12, pp.42–49, Dec. 2009.

[31] G. Fodor, E. Dahlman, G. Mildh, S. Parkvall, N. Reider, G. Miklós, and Z. Turányi, " Design Aspects of Network Assisted Device-to-Device Communications," IEEE Commun. Mag., vol.50, no.3, pp.170–177, Mar. 2012.

[32] A. Goldsmith, S.A. Jafar, I. Maric, and S. Srinivasa, "Breaking Spectrum Gridlock with Cognitive Radios: An Information Theoretic Perspective," IEEE Proc., vol. 97, no. 5, pp. 894–914, May 2009.

[33] S. Y. Lien, S. M. Cheng, S. Y. Shih, K. C. Chen, "Radio Resource Management for QoS Guarantees in Cyber-Physical Systems, " IEEE Trans. Parallel Distrib. Syst., vol.23, no.9, pp.1752–1761, Sep. 2012.

[34] D. Wu and R. Negi, "Effective capacity: a wireless link model for support of quality of service," IEEE Trans. Wireless Commun., vol. 12, no. 4, pp. 630–643, July 2003.

[35] C. S. Chang, "Stability, queue length, and delay of deterministic and stochastic queuing networks," IEEE Trans. Autom. Control, vol. 39, no. 5, pp. 913–931, May 1994.

[36] Qing Wang, Bongjun Ko, Kwang-Cheng Chen, Junsong Wang, Ting He, Yonghua Lin, Kang-won Lee, "Dynamic Spectrum Allocation under Cognitive Cellular Network for M2M Applications", IEEE Asilomar Conference, 2012.

[37] Y. Xiao, C. Yuen, L.A. DaSilva, K.C. Chen, "Spectrum Sharing for Device-to-Device Communications in Cellular Networks: A Game Theory Approach", to appear in the IEEE Dynamic Spectrum Access Networks (DySPAN), 2014.

[38] R. J.M. Vullers, R. van Schaihk, H.J. Visser, J. Penders, C. van Hoof, "Energy Harvesting for Autonomous Wireless Sensor Networks", IEEE Solid State Circuits Magazine, pp. 29–38, Spring 2010.

[39] K.C. Chen, R. Prasad, Cognitive Radio Networks, Wiley, 2009.

[40] I.F. Akyildiz, W.Y. Lee, K.R. Chowdhury, "CRAHNs: Cognitive Radio Ad Hoc Networks", Ad Hoc Networks, pp. 810–836, Volume 7, Issue 5, 2009.

[41] Y.C. Liang, K.C. Chen, J. Y. Li, P. Mahonen, "Cognitive Radio Networking and Communications: An Overview", IEEE Transaction on Vehicular Technology, vol. 60, no. 7, pp. 3386–3407, Sep 2011.

[42] O. Akan, O. Karli, O. Ergul, "Cognitive Radio Sensor Networks", IEEE Network, pp. 34–40, vol. 23, no. 4, 2009.
[43] Y. Zhang, et al., "Cognitive Machine-to-Machine Communications" Visions and Potentials for the Smart Grid", IEEE Network, pp. 6–13, May/June 2012
[44] M. Abolhasan, T. Wysocki, E. Dutkiewicz, "A Review of Routing Protocols for Mobile Ad Hoc Networks", Ad Hoc Networks, pp. 1–22, no. 2, 2004.
[45] M. Cesana, F. Cuomo, E. Ekici, "Routing in Cognitive Radio Networks: Challenges and Solutions", Ad Hoc Networks, pp. 228–248, vol. 9, no. 3, May 2011.
[46] G. Ganesan, J. Li, "Cooperative Spectrum Sensing in Cognitive Radio, Part I: Two-User Networks", IEEE Transactions on Wireless Communications, vol. 6, no. 6, pp. 2204–2213, June 2007.
[47] G. Ganesan, J. Li, "Cooperative Spectrum Sensing in Cognitive Radio, Part II: Multiuser Networks", IEEE Transactions on Wireless Communications, vol. 6, no. 6, pp. 2214–2222, June 2007.
[48] J. Unnikrishnan, V.V. Veeravalli, "Cooperative Sensing for Primary Detection in Cognitive Radio", IEEE Journal on Selected Topics in Signal Processing, vol. 2, no. 1, pp. 18–27, Feb. 2008.
[49] J. Jia, J. Zhang, Q. Zhang, "Cooperative Relay for Cognitive Radio Networks", IEEE INFOCOM, 2009.
[50] K.B. Letaief, W. Zhang, "Cooperative Communications for Cognitive Radio Networks", Proceeding of IEEE, pp. 878–893, vol. 97, no. 5, May 2009.
[51] E. Beres, R. Adve, "Selection Cooperation in Multi-Source Cooperative Networks", IEEE Tr. on Wireless Communications, pp. 118–127, vol. 7, no. 1, January 2008.
[52] K. C. Chen, B. K. Cetin, Y. C. Peng, N. Prasad, J. Wang, and S. Lee, "Routing for cognitive radio networks consisting of opportunistic links," (Wiley) Wireless Communications and Mobile Computing, 2010.
[53] K. C. Chen, P. Y. Chen, N. Prasad, Y. C. Liang, and S. Sun, "Trusted cognitive radio networking," (Wiley) Wireless Communications and Mobile Computing, 2010.
[54] J. Andrews, et al. "Rethinking Information Theory for Mobile Ad Hoc Networks", IEEE Communications Magazine, pp. 94–101, December 2008.
[55] W. Chen, et al. "Coding and Control for Communication Networks", Queuing Systems, vol. 63, pp. 195–216, 2009.

[56] A. Goldsmith, et al. "Beyond Shannon: The Quest for Fundamental Performance Limits of Wireless Ad Hoc Networks", IEEE Communications Magazine, pp. 195–205, May 2011.
[57] C. Cormio, K.R. Chowdhury, "A Survey on MAC Protocols for Cognitive Radio Networks", Ad Hoc Networks, pp. 1315–1329, Volume 7, Issue 7, 2009.
[58] T. Tucek, H. Arslan, "A Survey of Spectrum Sensing Algorithms for Cognitive Radio Applications", IEEE Communication Surveys and Tutorials, pp. 116–130, vol. 11, no. 1, 1Q, 2009.
[59] E. Axell, G. Leus, E. G. Larsson, H.V. Poor, "Spectrum Sensing for Cognitive Radio", IEEE Signal Processing Magazine, pp. 101–116, May 2012.
[60] K.C. Chen, S.Y. Tu, C.K. Yu, "Statistical Inference in Cognitive Radio Networks", ChinaCom, Xi-An, Aug 2009.
[61] C.K. Yu, Shimi Cheng, K.C. Chen, "Cognitive Radio Network Tomography", the special issue of Achievements and Road Ahead: The First Decade of Cognitive Radio IEEE Transactions on Vehicular Technology, vol. 59, no. 4, pp. 1980–1997, April, 2010.
[62] A.Coates, A.O. Hero III, R. Nowak,B. Yu, "Internet Tomography", IEEE Signal Processing Magazine, vol. 19, no. 3, pp. 47–65, May 2002.
[63] T.W. Chiang, K.C. Chen, "Synthetic Aperture Radar Construction of Spectrum Map for Cognitive Radio Networking", International Conference on Wireless Communications and Mobile Computing, Caen, France, 2010.
[64] S.Y. Shih, K.C. Chen, "Compressed Sensing Construction of Spectrum Map for routing in Cognitive Radio Networks", Wireless Communications and Mobile Computing, vol. 12, no. 18, pp. 1592–1607, December 2012.
[65] Y. Wang, Z. Tian, C. Feng, "Sparsity Order Estimation an Its Application in Compressive Spectrum Sensing for Cognitive Radio", IEEE Tr. on Wireless Communications, vol. 11, no. 6, pp. 2116–2125, June 2012.
[66] C.H. Huang, K.C. Chen, "Dual-Observation Time Division Spectrum Sensing for Cognitive Radios", IEEE Transaction on Vehicular Technology, vol. 60, no. 7, pp. 3712–3725, October 2011.
[67] J.K. Sreefharan, V. Sharma, "Spectrum Sensing via Universal Source Coding", IEEE GLOBECOM, 2012.
[68] Q. Zhao, L. Tong, A. Awami, Y. Chen, "Decnetralized Cognitive MAC for Opportunistic Spectrum Access in Ad hoc Networks: A POMDP

Framework", IEEE Journal on Selected Areas in Communications, vol. 25, no. 3, April 2007.
[69] L. Lai, H. El Gamal, H. Jiang, H.V. Poor, "Cognitive Medium Access: Exploration, Exploitation, and Competition", IEEE Tr. on Mobile Computing, vol. 10, no. 2, pp. 239–253, Feb 2011.
[70] J. Ai, A.A. Abouzeid, "Opportunistic Specrum Access based on a Constrained Multi-Armed Bandit Formulation", Journal of Communications and Networks, vol. 11, no. 2, 2009.
[71] K. Unnikrishnan, V.V. Veeravalli, "Algorithms for Dynamic Spectrum Access with Learning for Cognitive Radio", IEEE Tr. on Signal Processing, vol. 58, no. 2, pp. 750–760, February 2010.
[72] K.W. Choi, E. Hossain, "Opportunistic Access to Spectrum Holes between Packet Bursts: A Learning-Based Approach", IEEE Tr. on Wireless Communications, vol. 10, no. 8, pp. 2497–2509, August 2011.
[73] T.V. Nguyen, H. Shin, T.Q.S. Quek, M.Z. Win, "Sensing and probing Cardinalities for Active Cognitive Radios", IEEE Tr. on Signal Processing, vol. 60, no. 4, pp. 1833–1848, April 2012.
[74] M.Z. Win, P.C. Pinto, L.A. Sheep, "A Mathematical Theory of Network Interference and Its Applications", Proceeding of the IEEE, vol. 97, no. 2, pp. 205–230, Feb 2009.
[75] M. Haenggi, R.K. Ganti, "Interference in large Wireless Networks", Foundations and Trends in Networking, vol. 3, no. 2, pp. 127–248, 2008.
[76] M. Haenggi, J. Andrews, F. Baccelli, O. Dousse, and M. Franceschetti,"Stochastic geometry and random graphs for the analysis and design of wireless networks," IEEE J. Sel. Areas Commun., vol. 27, no. 7, pp. 1029–1046, Sep. 2009.
[77] M.O. Jackson, Social and Economic Networks, Princeton University Press, 2009.
[78] D. Easley, J. Kleinberg, Networks, Crowds, and Markets, Cambridge University Press, 2010.
[79] M.E.J. Newman, Networks: An Introduction, Oxford University Press, 2010.
[80] B. Hajek, T. Ephremides, "Information Theory and Communications Networks: An Unconsummated Union", IEEE Tr. On Information Theory, vol. 44, pp. 2416–2434, October 1998.
[81] P. Gupta, P.R. Kumar, "The Capacity of Wireless Networks", IEEE Tr. On Information Theory, vol. 46, no. 2, pp. 388–402, March 2000.

[82] F. Xue, P.R. Kumar, "Scaling Laws for Ad-Hoc Wireless Networks: An Information Theoretic Approach", NOW Foundations and Trends in Networking, vol. 1. no. 2, pp. 145–270, 2006.
[83] C. Xie, K.C. Chen, X. Wang, "To Hop or Not to Hop in Machine-to-Machine Communications", IEEE Wireless Communication and Networking Conference (WCNC), 2013.
[84] M. Franceschetti, O. Dousse, D.N.C. Tse, P. Thiran, "Closing the Gap in the Capacity of Wireless Networks via Percolation Theory", IEEE Tr. on Information Theory, vol. 53, no. 3, pp. 1009–1018, March 2007.
[85] X. Gong, T.P.S. Chandrashekhar, J. Zhang, H.V. Poor, "Opportunistic Cooperative Networking: To Relay or Not To Relay?", IEEE Journal on Selected Areas in Communications, vol. 30, no. 2, pp. 307–314, February 2012.
[86] P. Gupta, P.R. Kumar, "The Capacity of Wireless Networks", IEEE Tr. on Information Theory, vol. 46, no. 2, pp. 388–404, March 2000.
[87] S.P. Weber, X. Yang, J.G. Andrews, G. de Veciana, "Transmission Capacity of Wireless Ad hoc Networks with Constraints", IEEE Tr. on Information Theory, vol. 51, no. 12. Pp. 4091–4102, December 2005.
[88] S. Weber, J.G. Andrews, N. Jindal, "An Overview of Transmission Capacity of Wireless Networks", IEEE Tr. on Communications, vol. 58, no. 12, pp. 3593–3604, December 2010.
[89] W.C. Ao, K.C. Chen, "Upper Bound on Broadcast Transmission Capacity of Heterogeneous Wireless Ad Hoc Networks", IEEE Communications Letters, vol. 15, no. 11, pp. 1172–1174, Nov. 2011.
[90] W.C. Ao, K.C. Chen, "Bounds and Exact Mean Node Degree and Node Isolation Probability in Heterogeneous Wireless Ad Hoc Networks with General Fading", IEEE Transaction on Vehicular Technology, vol. 61, no. 5, pp. 2342–2348, June 2012.
[91] W.C. Ao, S.M. Cheng, K.C. Chen, "Phase Transition Diagram for Underlay Heterogeneous Cognitive Radio Networks", IEEE GLOBECOM, Miami, 2010.
[92] V.K.S. Shante, S. Kirkpatrick, "An introduction to Percolation Theory", Advances in Physics, 20: 85, pp. 325–357, May 1971.
[93] H. Zhang and J. Hou, "Asymptotic critical total power for k-connectivity of wireless networks" IEEE/ACM Trans. Netw., vol. 16, no. 2, pp. 347–358, Apr. 2008.
[94] X. Ta, G. Mao, and B. Anderson, "On the giant component of wireless multihop networks in the presence of shadowing", IEEE Trans. Veh. Technol., vol. 58, no. 9, pp. 5152–5163, Nov. 2009.

[95] W.C. Ao, K.C. Chen, "Percolation-based Connectivity of Multiple Cooperative Cognitive Radio Ad Hoc Networks", IEEE GLOBECOM, 2011.
[96] W.C. Ao, S.M. Cheng, K.C. Chen, "Connectivity of Multiple Cooperative Cognitive Radio Ad Hoc Networks", IEEE Journal on Selected Areas in Communications, vol. 30, no. 2, pp. 263–270, Feb 2012.
[97] W.C. Ao, K.C. Chen, "Cognitive Radio-Enabled Network-Based Cooperation: From a Connectivity Perspective", to appear in the IEEE Journal on Selected Areas in Communications, November 2012.
[98] Y. Sung, S. Mistra, L. Tong, A. Ephremides, "Cooperative Routing for Distributed Detection in Large Sensor Networks", IEEE Journal on Selected Areas in Communications, vol. 25, no. 2, pp. 471–483, February 2007.
[99] A.E. Khandani, J. Abounadi, E. Modiano, L. Zheng, "Cooperative Routing in Static Wireless Networks", IEEE Tr. on Communications, vol. 55, no. 11, pp. 2185–2192, November 2007.
[100] S.C. Lin, K.C. Chen, "Spectrum Aware Opportunistic Routing in Cognitive Radio Networks", IEEE GLOBECOM, Miami, 2010.
[101] C.K. Yu, K.C. Chen, "Spectrum Map Retrieval Using Cognitive Radio Network Tomography", IEEE GLOBECOM (Workshop on Recent Advances in Cognitive Communications and Networks), 2011.
[102] Q. Zhu, Z. Yuan, J.B. Song, Z. Han, T. Basar, "Interference Aware routing Game for Cognitive Radio Multi-hop Networks", IEEE Journal on Selected Areas in Communications, vol. 30, no. 10, pp. 2006–2015, November 2012.
[103] A. Nath, V.N. Ekambaram, A. Kumar, P.V. Kumar, "Theory and Algorithms for Hop-Count-Based Localization with Random Geometric Graph Models of Dense Sensor Networks", ACM Tr. on Sensor Networks, vol.8, no. 4, Article 35, September 2012.
[104] A.E. Khandani, J. Abounadi, E. Modiano, L. Zheng, "Reliability and Route Diversity in Wireless Networks", IEEE Tr. on Wireless Communications, vol. 7, no. 12, pp. 4772–4776, December 2008.
[105] R. W. Yeung, Information Theory and Network Coding, Springer, 2008.
[106] P.Y. Chen, W.C. Ao, K.C. Chen, "Rate-delay Enhanced Multipath Transmission Scheme via Network Coding in Multihop Networks", IEEE Communications Letters, vol. 16, no. 3, pp. 281–283, March 2012.

[107] W.C. Ao, P.Y. Chen, K.C. Chen, "Rate-reliability-delay Trade-off of Multipath Transmission Using Network Coding", IEEE Transaction on Vehicular Technology, vol. 61, no. 5, pp. 2336–2342, June 2012.

[108] J. Wang, J. Cho, S. Lee, K.C. Chen, Y.K. Lee, "Hop-Based Energy Aware Routing Algorithm for Wireless Sensor Networks", IEICE Transactions, vol E93, no. 2, pp.305–316, 2010.

[109] E. Dall'Anese, G.B. Giannakis, "Statistical Routing for Multihop Wireless Cognitive Networks", IEEE Journal on Selected Areas in Communications, vol. 30, no. 10, pp. 1983–1993, November 2012.

[110] E. Smith, "Large-Deviation Principles, Stochastic Effective Actions, Path Entropies, and the Structure and Meaning of Thermodynamic Descriptions", Reports on Progress in Physics, 74–046601, 2011.

[111] H.B. Chang, S.M. Cheng, S.Y. Lien, K.C. Chen, "Statistical Delay Control of Opportunistic Links in Cognitive Radio Networks", IEEE International Symposium on Personal Indoor Mobile Radio Communications, 2010.

[112] P.Y. Chen, S.M. Cheng, W.C. Ao, K.C. Chen, "Multi-path Routing with End-to-end Statistical QoS Provisioning in Underlay Cognitive Radio Networks", IEEE INFOCOM Workshop, Shanghai, 2011.

[113] A. Alshamrano, X. S. Shen, L. L. Xie, "QoS Provisioning for Heterogeneous Services in Cooperative Cognitive Radio Networks", IEEE Journal on Selected Areas in Communications, vol. 29, no. 4, pp. 819–820, April 2011.

[114] W.C. Ao, K.C. Chen, "End-to-End HARQ in Cognitive Radio Networks", IEEE Wireless Communications and Networking Conference, Sydney, 2010.

[115] W.C. Ao, K.C. Chen, "Error Control for Local Broadcasting in Heterogeneous Wireless Ad Hoc Networks", IEEE Transactions on Communications, vol. 61, no. 4, pp. 1510–1519, April 2013.

[116] I.W. Lai, C.H. Lee, K.C. Chen, "A Virtual MIMO Path-Time Code for Cognitive Ad Hoc Networks", IEEE Communications Letters, vol. 17, no. 1, pp. 4–7, January 2013.

[117] H. Li, Z. Han, "Socially Optimal Queuing Control in Cognitive Radio Networks Subject to Service Interruptions: To Queue or Not to Queue?", IEEE Tr. on Wireless Communications, vol. 10, no. 5, pp. 1656–1666, May 2011.

[118] P.Y. Chen, V.A. Karyotis, S. Papavassiliou, K.C. Chen, "Topology Control in Multi-channel Cognitive Radio Networks with Non-uniform Node

Arrangements", IEEE symposium on Computers and Communications, 2011.

[119] R. Lam, K.C. Chen, "Congestion Control for M2M Traffic with Heterogeneous Throughput Demands", IEEE Wireless Communications and Networks Conference (WCNC), 2013.

[120] S. Gunawardena, W. Zhuang, "Capacity Analysis and Call Admission Control in Distributed Cognitive Radio Networks", IEEE Tr. On Wireless Communications, vol. 10, no. 9, pp. 3110–3120, Sep. 2011.

[121] L. Gu, S.C. Lin, K.C. Chen, "Small-World Networks Empowered Large Machine-to-Machine Communications ", IEEE Wireless Communications and Networks Conference (WCNC), 2013.

[122] K.C. Chen, M. Chiang, H.V. Poor, "From Technological Networks to Social Networks", IEEE Journal on Selected Areas in Communications, vol. 31, no. 9, pp. 548–572, September 2013.

[123] J. M. Kleinberg, "Navigation in a Small World", Nature, vol. 406, pp. 845–845, August 24, 2000.

[124] H. Inaltekin, M. Chaing, H.V. Poor, "Average Message Delivery Time for Small World Networks in the Continuum Limit", IEEE Tr. on Information Theory, vol. 56, no. 9, pp. 4447–4470, Sep. 2010.

[125] S.C. Lin, L. Gu, K.C. Chen, "Providing Statistical QoS Guarantees in Large Cognitive Machine-to-Machine Networks", IEEE GLOBECOM (Machine-to-Machine Communications Workshop), 2012.

[126] M.D. Francesco, S.K. Das, G. Anastasi, "Data Collection in Wireless Sensor Networks with Mobile Elements: A Survey", ACM Tr. on Sensor Networks, vol. 8, no. 1, Article 7, August 2011.

[127] J.-F. Chamberland, V.V. Veeravalli, " Wireless Sensors in Distributed Detection Applications", IEEE Signal Processing Magazine, pp. 16–25, May 2007.

[128] A.D. Wood, et al., "Context-Aware Wireless Sensor Networks for Assisted Living and Residential Monitoring", IEEE Network, pp.26–33, July/August 2008.

[129] P. Kasirajan, C. Larsen, S. Jagannathan, "A New Data Aggregation Scheme via Adaptive Compression for Wireless Sensor Networks", ACM Tr. on Sensor Networks, vol. 9, no. 1, Article 5, November 2012.

[130] A. Giridhar, P.R. Kumar, "Computing and Communicating Functions over Sensor Networks', IEEE Journal on Selected Areas in Communications, vol. 23, no. 4, pp. 755–764, April 2005.

[131] E. Fasolo, M. Rossi, J. Widmer, M. Zorzi, "In-Network Aggregation Techniques for Wireless Sensor Networks: A Survey", IEEE Wireless Communications, pp. 70–87, April 2007.

[132] F. Li, "NASC: Network-Aware Source Coding for Wireless Broadcast Channels with multiple Sources", IEEE Vehicular Technology Conference – Fall, 2006.

[133] L.D. Davisson, "Universal Noiseless Coding", IEEE Tr. on Information Theory, vol. 19, no. 6, pp. 783–795, November 1973.

[134] T. Cui, L. Chen, T. Ho, S. Low, "Opportunistic Source Coding for Data Gathering in Wireless Sensor Networks", IEEE International Conference on Mobile Ad Hoc and Sensor Systems, 2007.

[135] K.C. Chen, F.M. Tseng, C.H. Lin, "In-Network Computations in M2M Communications for Wireless Robotics", Wireless Personal Communications, 2013.

[136] K.H. Peng, K.C. Chen, S.L. Huang, S.C. Hung, "Green Traffic Compression in Wireless Sensor Networks", IEEE Vehicular Technology Conference – Spring, 2014.

[137] S.C. Lin, K.C. Chen, "Improving Spectrum Efficiency via In-Network Computations in Cognitive Radio sensor Networks", to appear in the IEEE Tr. On Wireless Communications.

[138] M. Chiang, S.H. Low, A.R. Calderbank, J.C. Doyle, "Layering as Optimization Decomposition: A Mathematical Theory of Network Architectures", Proceeding of the IEEE, vol. 95, no. 1, pp. 255–312, January 2007.

[139] D. Liau, K.C. Chen, S.M. Cheng, "A Predator-Prey Model for Dynamics of Cognitive Radios", IEEE Communications Letters, vol. 17, no. 3, pp. 467–470, March 2013.

[140] Y. Zhang, R. Yu, S. Xie, W. Yao, Y. Xiao, and M. Guizani, "Home M2M networks: Architectures, standards, and QoS improvement," IEEE Commun. Mag., vol. 49, no. 4, pp. 44–52, Apr. 2011.

[141] X. Li, et al., "Smart Community: An Internet of Things Application", IEEE Communications Magazine, pp. 68–75, November 2011.

[142] J.H. Porter, P.C. Hanson, C C. Lin, "Staying Afloat in the Sensor Data Deluge", Trends in Ecology and Evolution, vol. 27, no. 2, pp. 121–129, February 2012.

[143] K.C. Chen, "Machine-to-Machine Communications for Healthcare", Journal of Computing Science and Engineering, vol. 6, no. 2, June 2012.

[144] H. Hartenstein, K.P. Laberteaux, "A Tutorial Survey on Vehicular Ad Hoc Networks", vol. 46, no. 6, pp. 164–171, June 2008.

[145] I. Mezei, V. Malbaa, I. Stojmenovic, "Robot to Robot", IEEE Robotics & Automation Magazine, pp. 63–69, December 2010.
[146] V. Sharma, U. Mukherji, V. Joseph, S. Gupta, "Optimal Energy Management Policies for Energy Harvest Senor Nodes", IEEE Tr. on Wireless Communications, vol. 9, no. 4, pp. 1326–1336, April 2010.
[147] O. Ozel, K. Tutuncuoglu, J. Yang, S. Ulukus, A. Yenar, "Transmission with Energy Harvesting Nodes in Fading Wireless Channels: Optimal Policies", IEEE Journal on Selected Areas in Communications, vol. 29, no. 8, pp. 1732–1743, September 2011.
[148] J. Yang, S. Ulukus, "Optimal Packet Scheduling in an Energy Harvesting Communication System", IEEE Trans. on Communications, vol. 60, no. 1, pp. 220–230, January 2012.
[149] J. Yang, O. Ozel, S. Ulukus, "Broadcasting with an Energy Harvesting Rechargeable Transmitter", IEEE Tr. on Wireless Communications, vol. 11, no. 2, pp. 571–583, February 2012.
[150] K. Tutuncuoglu, A. Yenar, "Optimum Transmission Policies for Battery Limited Energy Harvesting Nodes", IEEE Tr. on Wireless Communications, vol. 11, no. 3, pp. 1180–1189, March 2012.
[151] P. Blasco, D. Gunduz, M. Dohler, "A Learning Theoretic Approach to Energy Harvesting Communication System Optimization", IEEE GLOBECOM Workshop (IWM2MC), 2012.
[152] P. McDaniel, S. McLaughlin, "Security and privacy Challenges in the Smart Grid", IEEE Security & Privacy, vol. 7, no. 3, pp. 75–77, May-June, 2009.
[153] J. Krumm, "A Survey of Computational Location Privacy", Pervasive Ubiquitous Computing, vol.13, pp. 391–399, 2009.
[154] L. Sankar, S. Raj Rajagopalan, H.V. Poor, "A Theory of Utility and Privacy of Data Sources", IEEE International Symposium on Information Theory, 2010.
[155] P. Traynor, W. Enck, P. McDaniel, T. La Porta, "Mitigating Attacks on Open Functionality in SMS-Capable Cellular Networks", IEEE/ACM Tr. on Networking, vol. 17, no. 1, pp. 40–53, February 2009.
[156] S.M. Cheng, P.Y. Chen, K.C. Chen, "Ecology of Cognitive Radio Ad Hoc Networks", IEEE Communications Letters, vol. 15, no. 7, pp. 764–766, July, 2011.
[157] P.Y. Chen, S.M. Cheng, K.C. Chen, "Smart Attacks in Smart Grid Communication Networks", IEEE Communications Magazine, vol. 50, no. 8, pp. 24–29, August 2012.

[158] P.Y. Chen, K.C. Chen, "Intentional Attack and Fusion-based Defense Strategy in Complex Networks", IEEE GLOBECOM, 2011.
[159] L.B. Le, E. Hossain, "Resource Allocation for Spectrum Underlay in Cognitive Radio Networks", IEEE Tr. on Wireless Communications, vol. 7, no. 12, pp. 5306–5315, December 2008.
[160] R. Zhang, Y.C. Liang, S. Cui, "Dynamic Resource Allocation in Cognitive Radio Networks", IEEE Signal Processing Magazine, vol.27, no. 3, pp. 102–114, May 2010.
[161] Z. Shu, J. Zhou, Y. Yang, H. Sharif, Y. Qian, "Network Coding-aware Channel Allocation and Routing in Cognitive Radio Networks", IEEE GLOEBCOM, 2012.
[162] H.B. Chang, K.C. Chen, "Auction Based Spectrum Management of Cognitive Radio Networks", the special issue of Achievements and Road Ahead: The First Decade of Cognitive Radio, IEEE Transactions on Vehicular Technology, vol. 59, no. 4, pp. 1923–1935, April 2010.
[163] H.B. Chang, K.C. Chen, "Cooperative Spectrum Sharing Economy for Heterogeneous Wireless Networks", IEEE GLOBECOM (Multicell Cooperation Workshop), 2011
[164] P.Y. Chen, W.C. Ao, S.C. Lin, K.C. Chen, "Reciprocal Spectrum Sharing Game and Mechanism in Cellular Systems with Cognitive Radio Users", IEEE GLOBECOM (Workshop on Recent Advances in Cognitive Communications and Networks), 2011.
[165] N.K. Pratas, H. Thomsen, C. Stefanovic, P. Popovski, "Coded-Expanded Random Access for Machine-Type Communications', IEEE GLOBE-COM Workshop (IWM2MC), 2012.
[166] R. Olfati-Saber, J. Fax, and R. Murray, "Consensus and cooperation in networked multi-agent systems," Proceedings of the IEEE, vol. 95, no. 1, pp. 215 –233, Jan. 2007.
[167] Y.Y. Lin, K.C. Chen, "Asynchronous Dynamics Spectrum Access", IEEE Transaction on Vehicular Technology, vol.61, no.1, pp. 222–236, Jan 2012.
[168] B. Treister, A. Batra, K.C. Chen, O. Eliezer, "Adaptive Frequency Hopping: A Non-Collaborative Coexistence Mechanism", IEEE P802.15–01/252r0, 2001.
[169] C.K. Liang, K.C. Chen, "A Green Software-Defined Communication Processor for Dynamic Spectrum Access", IEEE International Symposium on Personal Indoor Mobile Radio Communications, 2010.

[170] R. Di Pietro, D. Ma, C. Soriente, G. Tsudik, "Self-Healing in Unattended Wireless Sensor Networks", ACM Tr. on Sensor Networks, vol. 9, no. 1, Article 7, November 2012.
[171] A. Bahga, V.K. Madisetti, "Analyzing Massive Machine Maintenance Data in a Computing Cloud", IEEE Tr. On Parallel and Distributed Systems, vol. 23, no. 10, pp. 1831–1843, Oct. 2012.
[172] L. Mottola, G.P. Picco, "Programming Wireless Sensor Networks: Fundamental Concepts and State of the Art", ACM Computing Survey, vol. 43, no. 3, Article 19, April 2011.
[173] J. Sztipanovits, et al., "Toward a Science of Cyber-Physical System Integration", Proceeding of the IEEE, vol. 100, no. 1, pp. 29–44, January 2012.

Biographies

Kwang-Cheng Chen (M'89-SM'94-F'07) received the B.S. from the National Taiwan University in 1983, and the M.S. and Ph.D from the University of Maryland, College Park, United States, in 1987 and 1989, all in electrical engineering. From 1987 to 1998, Dr. Chen worked with SSE, COMSAT, IBM Thomas J. Watson Research Center, and National TsingHua University, in mobile communications and networks. Since 1998, Dr. Chen has been with National Taiwan University, Taipei, Taiwan, ROC, and is the *Distinguished Professor and Associate Dean* for academic affairs in the College of Electrical Engineering and Computer Science, National Taiwan University. He has been actively involving in the organization of various IEEE conferences as General/TPC chair/co-chair, and has served in editorships with a few IEEE journals and many international journals and in various positions with IEEE and various societies. Dr. Chen also actively participates in and has contributed essential technology to various IEEE 802, Bluetooth, and 3GPP wireless standards. He has authored and co-authored over 250 technical papers and more than 20 granted US patents. He co-edited (with R. DeMarca) the book *Mobile WiMAX* published by Wiley in 2008, and authored the book *Principles of Communications* published by River in 2009, and co-authored (with R. Prasad) another book *Cognitive Radio Networks* published by Wiley in 2009. Dr. Chen is an IEEE Fellow and has received a number of awards including the *2011 IEEE COMSOC WTC Recognition Award* and has co-authored a few award-winning papers published in the IEEE ComSoc journals and conferences. Dr. Chen's research interests include wireless communications and network science.

Shao-Yu Lien received his B.S. degree from the Department of Electrical Engineering, National Taiwan Ocean University, Taiwan, in 2004, the M.S. degree from the Institute of Computer and Communication Engineering, National Cheng Kung University, Taiwan, in 2006, and the Ph.D degree from the Graduate Institute of Communication Engineering, National Taiwan University, Taiwan, in 2011. After the one year military service as a second lieutenant platoon leader, he joined the Communication Research Center, National Taiwan University, as the post-doctorial research fellow. Since 2013, he has been with the Department of Electronic Engineering, National Formosa University, Taiwan, as an assistant professor. Prof. Lien received a number of prestigious recognitions, including IEEE ICC 2010 Best Paper Award and URSI AP-RASC 2013 Young Scientist Award. His research interests lie in optimization techniques for networks and communication systems. Recently, focuses are particularly on cyber-physical systems (including machine-to-machine communications, smart grid communications, cloud computing networks), 4G/5G communication networks (including cognitive radio technology, device-to-device communications, heterogeneous networks), and network sciences (including social networks, complex networks, large-scale networks).

7

Maximizing Throughput in Chip to Chip Communications

Hristomir Yordanov[1], Albena Mihovska[2], Vladimir Poulkov[1] and Ramjee Prasad[2]

[1]Technical University Sofia, Bulgaria
[2]CTIF (Aalborg University,Denmark)

7.1 Introduction

The rapid increase of the computational power of the integrated circuits in the last decade has created opportunities for data collection and processing from all possible sources. Such enormous information transfer requires communication channels with maximum possible capacity. The growing number of devices makes the implementation of wired links more difficult. For example routing a circuit board with a couple of FPGAs, several fast ADCs, and some interface chips can take months of work or even more. Additionally there are many applications where connecting cables turn out to be impractical or even impossible to use. And finally, there are some cases where the wired interface is not fast enough. For instance a parallel data bus in a PC cannot provide the data from the RAM to the CPU with sufficient speed. This bottleneck is currently solved by caching data in the CPU.

The analogue circuit technology has also enjoyed rapid development and allows for fabrication of high-speed wireless data links. The cheap and reliable CMOS process supports generators, amplifiers, modulators and detectors way in the millimetre-wave range and wideband transceivers in the 57 – 64 GHz are available. The antennas operating at those frequencies are of very small dimension. Additionally, the silicon substrate of the integrated circuit has a very high dielectric permittivity ($\varepsilon_r = 11.9$ @ 10 GHz), which renders the wavelength and thus the antenna size even smaller. For instance, the wavelength in silicon at 60 GHz is 1.44 mm. This allows for integrating

Convergence of Communications, Navigation, Sensing and Services, 181–200.

the antennas on the chip and thus developing fully integrated and highly miniaturized transceivers, which can be used for chip-to-chip communication, wireless sensor systems, automation and control systems, etc.

This chapter describes the challenges in maximizing the throughput of a fully integrated chip to chip communication channel. The challenges in manufacturing an efficient antenna integrated on a silicon chip are presented. An area-efficient solution, based on sharing of metallization structures between the antenna and the CMOS circuitry is described in detail, including study of the chip substrate effects on the antenna parameters. The wireless communication channels between chips have been evaluated and a review of various techniques for increasing the data throughput is included.

7.2 Antenna Implementation and Challenges

The development of Monolithic Microwave Integrated Circuits (MMICS) based on gallium arsenide substrate started in 1976 (Pengelly & Turner, 1976). It allowed the fabrication of on-chip millimetre wave devices. Low profile antennas like printed or dielectric antennas could also be manufactured using this technology and the first successful implementations were reported in the beginning of the 1980's (Yao, & Blumenstock, 1982), (Jain & Bansal, 1984). These designs exploited the low dielectric loss of the GaAs substrate at high frequencies to produce antennas of small dimensions and high radiation efficiency. There have been also successful antenna fabrications on silicon substrate. An overview of various printed patch antennas and arrays can be found in (Russer, 1998). The standard CMOS technology requires silicon with low resistivity and therefore high dielectric losses, therefore special care has to be taken to minimize those losses, like using high-resistivity Si or thinning the substrate. The integrated antennas have been used mostly in anti-collision radar systems.

The integration of antennas on silicon substrate drew attention in the late 1990's. By that time the CMOS technology roadmap was predicting the rise of digital clock speeds up in the gigahertz range. There was a concern among researchers that so high switching rates could be a reason for a dispersion of the clock signal along the line within the chip and as a result there could be a failure in the synchronization of the different elements of the circuit. A solution for a wireless clock distribution is possible (Kim, Floyd, & Mehta, 1997), using dipole and loop antennas, operating at 20 GHz. The potential for wireless chip-to-chip communication was quickly recognized and experiments were performed for measuring the channel gain of two antennas manufactured on the same chip (Kim, 1998). The obtained results were in the range of −50 dB for distances between the antennas in the range of 10 mm.

The proposed antennas suffer three major drawbacks. First, the p-doped silicon used in the standard CMOS process has a resistivity ρ in the range of 5 – 20 ?cm. This is very low for operation at microwave frequencies, as it leads to high dielectric losses. Therefore one has to either use high-resistivity silicone with ? = 2 – 5 k?cm, or to use a very thin substrate. Both techniques have their drawbacks. High-resistivity silicon is prone to formation of surface channels at the interfaces between the substrate and the oxide and between the substrate and metallisation, thus significantly degrading the excellent RF qualities of the material (Spirito, et al., 2005). It has to be passivated in order to prevent this formation. Additionally low-resistivity wells have to be manufactured for the high density transistor integration. This requires at least two fabrication steps beyond the standard CMOS process. The alternative technique of thinning the substrate can be more complicated and expensive, as the chips must be thinned from the back side.

The second drawback of the proposed antennas is that they require dedicated chip area. Any conducting structure in the vicinity of the antenna can significantly influence its radiation characteristics. Therefore a certain clearance must be provided. This is difficult to do in standard CMOS technology, where chip area is very expensive.

Finally, the dipole antennas are very narrow-band by nature and provide a small channel capacity at lower center frequencies. Manufacturing wideband antennas is impractical, because they require even greater chip area than the dipole and the loop antenna, as is the case for a fractal wideband antenna (Kikkawa, Kimoto, & Watanabe, 2005). This problem solves itself with the development of the CMOS technology, which allows ever greater operating frequency, which, in turn, means higher absolute bandwidth and therefore higher communication rates.

Different techniques have been proposed to overcome some of these challenges. Mendes *et. Al, 2004* have proposed a patch antenna using the CMOS circuitry as a ground plane, a dedicated high resistivity polycrystalline silicon as dielectric and dedicated metallisation for the patch. In this solution there is no electromagnetic field penetrating the lossy CMOS substrate, as the circuit metallisation serves as a shield, hence a high radiation efficiency. Additionally the antenna is stacked on the top of the integrated circuit, therefore no additional chip area is required. This solution is still narrowband and operating at relatively low frequencies, which does not allow for high data rates.

The latest CMOS technology generations allow building circuits operating in the millimetre wave range. This suggests reduced antenna dimensions. Pan and Capolino have suggested building a slot antenna backed with a resonating cavity, resonating at 140 GHz (Pan & Capolino, 2011). The cavity is to be embedded in the CMOS metallisation layers. The dielectric material filling the cavity is silicon dioxide, which has low losses and the obtained antenna has good efficiency. The antenna occupies an area of $1.2 \times 0.6\ \mathrm{mm}^2$ though , which could be considered significant.

7.3 Area-Efficient Antennas

The clock frequency of the 22 nm CMOS technology, which is the latest generation at the time of writing, supports clock speeds of about 4 GHz. The wireless inter-chip communication channels operate best in the 60 GHz band. This large difference in the operating frequency suggests that the two systems can function largely independent from each other. Therefore they can share on-chip structures. For instance, the ground supply plane of the CMOS circuit, which is usually manufactured in the top metal layer of the chip can be used for building antenna electrodes (Yordanov & Russer, 2008). In this case there is no additional chip area required by the antenna.

Another on-chip structure that can be used as a radiator are the bonding wires (Russer, Mukhtar, Wane, Bajon, & Russer, 2012). This approach requires designing a feeding network that couples both the RF and the digital signal through the wire, but once again no additional chip area is required.

This section will cover the various aspects of antennas sharing metallisation with the CMOS circuitry ground plane. The problems that will be covered are the antenna current distribution and radiation, interference with the digital circuitry and interconnects, and substrate losses.

7.3.1 Antenna Structure and Radiation Mode

A cross-section view of the integrated antenna using the circuit ground plane as electrodes is presented in Figure 7.1. For clarity the scale of the figure has been adjusted. The antenna is fabricated either on high-resistivity (at least 1 k?cm) substrate with thickness of 675 μm, or on thin low-resistivity substrate, as discussed in the previous section. The active devices are fabricated on the top of the substrate. Several metallisation layers follow, where the interconnects are fabricated. The top metallisation layer holds the ground supply plane. This plane is cut into patches and an RF generator is connected across the resulting gaps, exciting the antenna.

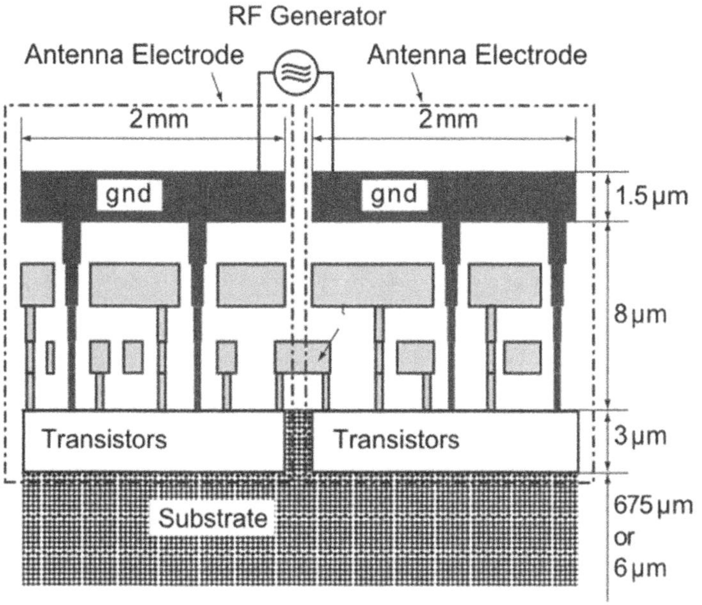

Figure 7.1 Detailed view of the cross-section of the integrated on-chip antenna, using the ground planes as antenna electrodes. The separated areas of the ground planes have to be connected to each other using inductive connections. The RF generator is also integrated in the CMOS circuit

Since the patches serve as ground planes also for the power supply for the CMOS circuitry underneath them, a low-frequency galvanic connection must be provided between them. This could be done using block inductors, providing a DC connection between the patches.

The antenna structure consists of two or more patches, separated by narrow gaps. If the length of the gaps is comparable with the wavelength, they can be treated as slot lines. If these lines are terminated with an open or short-circuit, a standing wave pattern will be formed along the line. This standing wave provides a time-varying electric polarisation, which is a source of radiation. Figure 7.2 shows the current distribution in a four-patch antenna configuration.

The guided wavelength in the slot line is given by (Cohn, 1969)

$$\lambda_g = \frac{\lambda_0}{\sqrt{\frac{1}{2}(\varepsilon_{r,Si}+1)}},$$

where λ_0 is the free-space wavelength and $\varepsilon_{r,Si}$ is the relative permittivity of the substrate. For a carrier frequency $f = 66$ GHz the corresponding

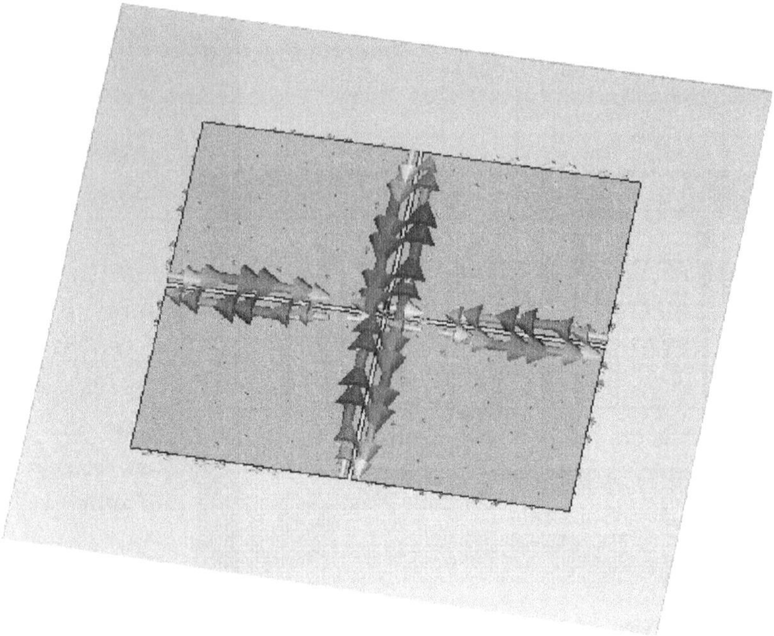

Figure 7.2 Current Distribution in an 2 × 2 Patch Antenna Configuration

wavelength is $\lambda_g = 1.8$ mm. Therefore an open-circuited slot with a length of 1.35 mm becomes a $3/4\lambda_g$resonator. Simulations of a two-patch structure, presented in Figure 7.3 show that the length of the slot should be a bit shorter, namely 1.1 mm, accounting for the effective slot elongation Δl due to the stray capacitance of the open slot line end. The current distribution, plotted in the figure, shows the variation in the current density along the gap.

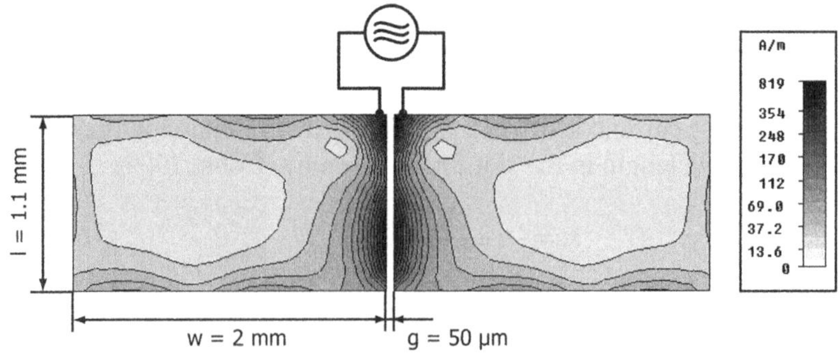

Figure 7.3 Top View and Current Distribution of a Two Patch Antenna, Operating at 66GHz

The input impedance of the antenna depends on the gap length, gap width, and substrate permittivity and structure. However, the substrate is defined by the CMOS technology process, and the gap length is defined by the antenna resonant frequency. The only parameter the designer can tune in order to match the antenna to a given load impedance is the gap width. Figure 7.4 presents the real and imaginary part of the antenna input impedance vs. frequency for different values of the gap width g ranging from 20 to 100 μm. It can be seen that increasing the gap width increases the real part of the input impedance. This could be explained by the fact that the gap width defines the characteristic capacitance of the slot line, where the radiating standing wave is formed. Increasing the gap means decreasing the capacitance, and,

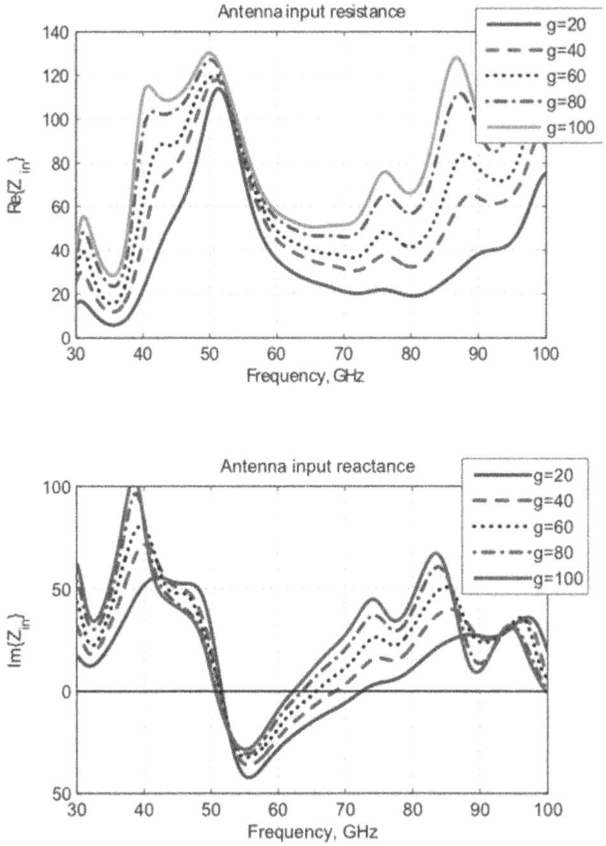

Figure 7.4 Real (top) and imaginary (bottom) part of the input impedance of the two patch antenna as given in Figure 7.3 for different values of the gap width g (in μm) vs. frequency

since the characteristic impedance of the line is

$$Z_C = \sqrt{\frac{L}{C}},$$

the line impedance is increased. The variation of the gap width changes the resonant frequency of the antenna, as can be seen in Figure 7.4, because it changes the effective line elongation. Increased gap means greater elongation Δl and therefore lower resonant frequency.

From the reactance plot (Figure 7.4) it can be seen that for $g = 50$ µm the antenna has two resonances in the selected frequency band. The first one occurs at about 51 GHz and the second one at 66 GHz. In the vicinity of the second resonance the real part of the input impedance is flat in a rather wide range, which means that the antenna has a higher bandwidth for that mode.

7.3.2 Interference Issues

There are two types of interference problems associated with monolithic antennas, using the circuit ground planes. The first one is the interference of the digital interconnects in the antenna mode. Under the antenna electrodes there are additional metallisation structures, which connect the active devices (see Figure 7.1), that have not been considered in the investigation of the antenna mode in the previous section. The second issue is the electromagnetic interference, which may cause two effects; antenna currents could couple to the transistors and thus degrade the functionality of the digital circuit; and transistor noise could couple to the antenna and increase its microwave noise levels, thus degrading the signal-to-noise ratio of a received signal.

It can easily be shown that the antenna current can not degrade the bit error rate of the digital circuitry. The antenna operates in the millimetre wave range, say in the 57 – 64 GHz band. The MOS transistors work at about 4 GHz clock frequency in the latest technology generations. The MOSFETs are known to have low-pass characteristics, that is any antenna current will be blocked by the first transistor it enters and will not propagate to the digital outputs of the device. Experiments have shown that irradiating a DRAM memory chip with an electromagnetic wave with a frequency of 20 GHz and 120 mW radiated power increase the memory bit error rate negligibly; much less than a 10 degrees change in the chip core temperature does (Bohorquez, 2004).

The second source of electromagnetic interference, the induction of noise in the receiver antenna due to the switching currents of the transistors, has not been studied yet. Preliminary work suggests that the induced noise is negligible, due to the great frequency difference between the circuit switching rates and the receiver band.

There are two types of interconnects available on the chip with respect to the integrated antenna type of interest in this section. There are interconnects that are situated underneath the antenna patches and there are interconnects that bridge the gap between the patches.

The length of the interconnects underneath the patches is on average much smaller than the wavelength of the antenna field. From the theory of dielectric materials we learn that the dielectric can be modelled as perfectly conducting spheres not touching each other, placed in vacuum. Consider a parallel plate capacitor in vacuum. The capacitance C is connected to the charge on each plate Q and the voltage across the terminals V via

$$C = \frac{Q}{V}.$$

If we place a dielectric material between the plates, as shown in Figure 7.5, the capacitance rises in proportion to the dielectric permittivity of that material. Since the charge remains the same, this means that the voltage has dropped. If we place a conducting block between the capacitor plates so that the block does not touch the plates, the charge of the capacitor induces an opposite charge in the metal. Therefore the electric field spreads across the gap between the plates and the metal, whereas in the free-space capacitor it spreads between the two plates. Since the charge is the same in both cases, the electric field density is the same. The voltage is the electric field density times the distance, therefore placing a metal block between the plates reduces the voltage and therefore increases the capacitance, i.e. it has the same effect as the dielectric material. This suggests that we can model the dielectric material as an array of conducting spheres in vacuum. Therefore, we can also model the volume under

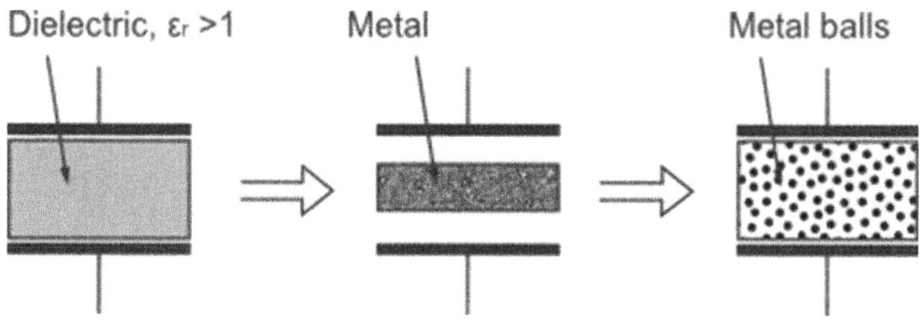

Figure 7.5 A dielectric material with $\varepsilon_r > 1$ is equivalent to an array of small conducting elements

the antenna patches where the short interconnects are as dielectric material with permittivity

$$\varepsilon_r = \frac{\varepsilon_{r,\mathrm{SiO_2}}}{1-\eta},$$

where $\varepsilon_{r,\mathrm{SiO_2}}$ is the permittivity of silicon dioxide, which embeds the interconnects, and η is the proportion of the volume, occupied by the metallisation. Both simulation and measurement confirm the validity of the model.

The interconnects that bridge the areas between different patches are parallel to the electric field in the antenna. Figure 7.6a, shows a four wire bus across the antenna gap. The field distribution of the antenna in the absence of the wires is shown in Figure 7.6b. The antenna is excited at one side of the gap at the bottom side of the figure. When the interconnects are present, however, the field distribution changes and there is no field beyond the wires, because they effectively short-circuit the slot, as shown in Figure 7.6c.

The measurements confirm these results. Figure 7.7a shows a photograph of the fabricated antenna with feeding balun on the top of the picture and interconnects across the gap. The return loss of the antenna in the absence and for two various positions of the interconnects is plotted in Figure 7.7b. The interconnects alter the field distribution and therefore the input impedance and the return loss of the antenna.

The areas under the antenna patches are digital CMOS circuits and there must be a possibility to draw interconnects between them in order to provide suitable communication. An elegant way to do this is to short-circuit the slot

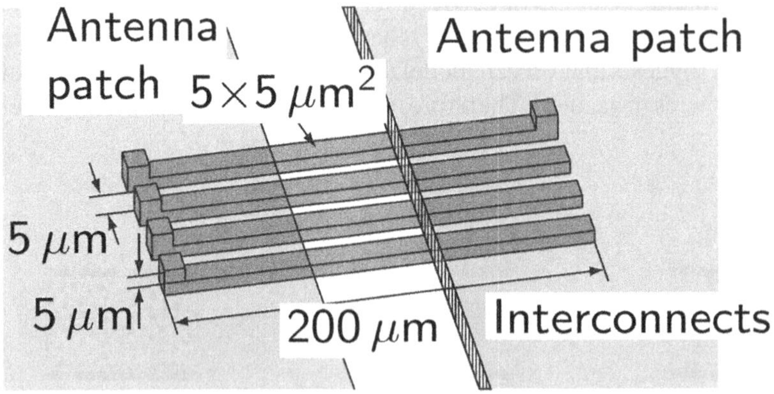

Figure 7.6 Geometry of the investigated interconnect. Electric field distribution in the slot between the patches in the absence (b) and in the presence (c) of an interconnection bus

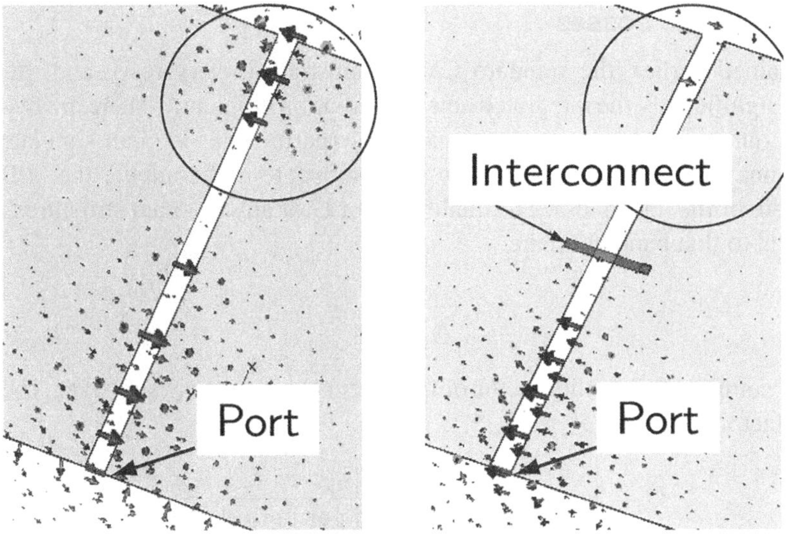

Figure 7.7 Setup (a) and measurement results (b) for estimation of the antenna mode distortion introduced by cross-patch interconnects for various positions of the interconnects relative to the slot length

in a position where the electric field is zero. This will not alter the antenna field and will provide a shielded section for the bridging interconnects. A die photograph and the measured return loss of such an antenna is provided in Figure 7.8. The figure also shows that there is no interference of the antenna mode due to the shielded interconnects, because the return loss is the same in the presence and in the absence of the wires.

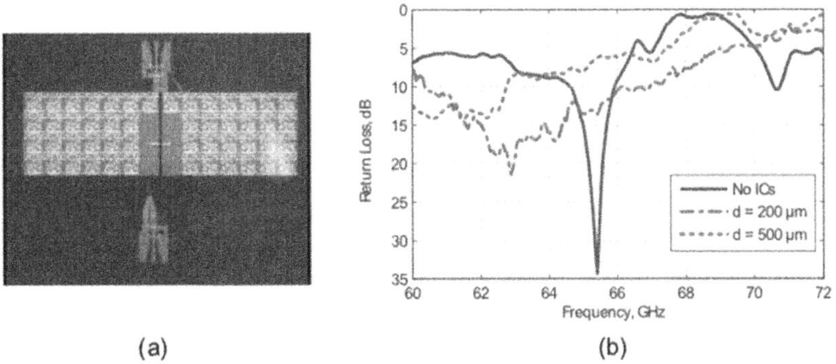

Figure 7.8 Short-circuited Slot Antenna Return Loss (b) in the Presence (a) And in the Absence of Shielded Interconnects

7.3.3 Substrate Losses

As explained earlier, the standard CMOS substrate is very lossy and this reduces significantly the integrated antenna radiation efficiency. There are two ways to quantify this effect: numerical and experimental. We can simulate the antenna on a computer with a lossless substrate and compute the 3dB bandwidth of the return loss. The quality factor Q of any resonant structure is reciprocal to that bandwidth, i.e.

$$Q = \frac{1}{\Delta f_{3\text{ dB}}}\,.$$

We also compute the quality factor of the antenna with a lossy substrate. The quality factor is defined as

$$Q = 2\pi\frac{\text{Stored energy}}{\text{Energy dissipated per period}}\,.$$

The antenna dissipates energy either by radiating it as electromagnetic field, which is desired, or due to losses in its metal and the dielectric components, which is undesired. Comparing the quality factor simulations of the lossless and the lossy antenna we can estimate how much energy is dissipated due to substrate losses, and the efficiency of the antenna, which is given by

$$\text{Efficiency} = \frac{\text{Radiated power}}{\text{Input power}} = \frac{\text{Radiated power}}{\text{Radiated power} + \text{Power Losses}}\,.$$

Unfortunately this method could not be applied experimentally, because we cannot fabricate a perfectly lossless antenna in order to measure its 3 dB bandwidth. What we can do is fabricate antennas on high-resistivity substrate and on thin substrate, measure the insertion loss of a link between two antennas and compare that to estimate which of the two solutions provides smaller losses.

The fabrication of antennas on high resistivity silicon follows the standard CMOS process, provided that some adjustments are made for the reduced thermal conductivity of the substrate. The fabrication and measurement of such prototypes is described in literature (Yordanov & Russer, 2010).

There are several methods available for manufacturing thin substrates. The one that provides the thinnest dice is a high precision method, called ChipFilm^{TM} (Burghartz, Appel, Rempp, & Zimmermann, 2009), developed at the Institute for Microelectronics Stuttgart (IMS CHIPS), Germany, which allows for manufacturing chips with substrate thickness as low as 6 μm. In ChipFilm^{TM} technology a thin Si membrane is created and firmly attached to a conventional bulk silicon wafer by vertical silicon micro-anchors, which in the end are controllably fractured to make the thin chips detachable. The antennas can be integrated on the thin Si membrane wafers as part of the last metallization layer of the available 0.5 μm mixed-signal CMOS process. A detailed description of the fabrication and measurement of antennas on ChipFilm^{TM} is presented in (Yordanov & Angelopoulos, 2013).

The numerical investigation of the two antenna types shows that the high-resistivity substrate antennas have an efficiency of about 90%, whereas antennas built on 16 μm standard CMOS substrate have an efficiency of about 60%. The superiority of high-resistivity based antennas is backed by the experimental set-up, measuring the channel loss of a link between two antennas of the same type, as shown in Figure 7.19, where the loss is plotted for the two antenna types and for different mutual antenna positioning versus the distance between the antennas. The solid line represents the insertion loss for high-resistivity based antennas, placed in each other's direction of maximum radiation, the dotted line is for the same antennas, placed in each other's direction of minimum radiation, and the dashed-dotted line is for ChipFilm^{TM} based antennas, placed in each other's direction of maximum radiation. It is evident that the thin substrate based antennas exhibit more than 20 dB insertion loss more than the high-resistivity silicon antennas.

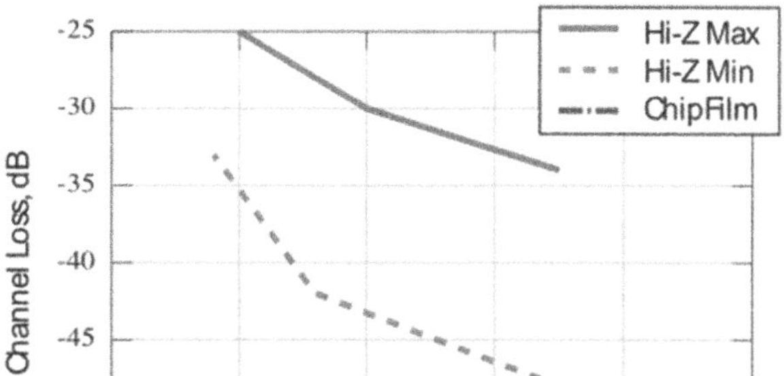

Figure 7.9 Insertion Loss of a Link Between two Integrated Antennas of the Same Type

7.4 Chip-to-Chip Communication

The measurement and simulation results allow the computation of the channel capacity of a wireless link with two high-impedance based antennas. It can be computed (Ivrlac & Nossek, 2010) using the representation of the channel as a 2-port. Using the impedance matrix representation of the 2-port we can compute the channel transmission function as

$$h(f) = \frac{Z_{21}}{\sqrt{\Re\{Z_{11}\}\Re\{Z_{22}\}}}.$$

The transmission function $h(f)$ is discrete in frequency, because $Z(f)$ is available in $n+1$ discrete data points, obtained by simulation or measurement. If the lowest frequency point for which the channel function is computed is $f_0$and the points are equally spaced over the bandwidth *B*, we can write the function $h(f)$ as the following series

$$h_0 = h(f_0), \, h_1 = h(f_0 + B), \, h_2 = h(f_0 + 2B), \, \cdots, \, h_n = h(f_0 + nB).$$

Then the channel throughput is given by

$$S = \sum_{k=0}^{n} B\log_2 \left(1 + \frac{P_k |h_k|}{\sigma^2}\right),$$

where $P_k$is the power radiated in each frequency range

$$f_0 + kB < f < f_0 + (k+1)B.$$

The noise power σ^2 is given by

$$\sigma^2 = 4kT_A B 10^{NF/10},$$

where T_A is the antenna noise temperature and NF is the receiver noise figure. We maximize the throughput *S* by using such values of the transmitted power, that the channel is most effectively used. In this way we get the obtainable channel capacity.

A plot for the capacity of a chip-to-chip channel with open-circuited slot antennas versus the transmission power is given in Figure 7.10, where the antenna temperature has been assumed to be $T_A = 300°$K and the receiver noise figure is $NF = 2$ dB. The distance between the antennas is 35 mm. Figure 7.11 shows the channel usage versus frequency.

The achievable channel capacity with 1 dBm transmission power is approximately 106.2 Gb/s. This is an overestimation of the achievable data rate, because the transmitter is assumed to have ideal transmission power density distribution.

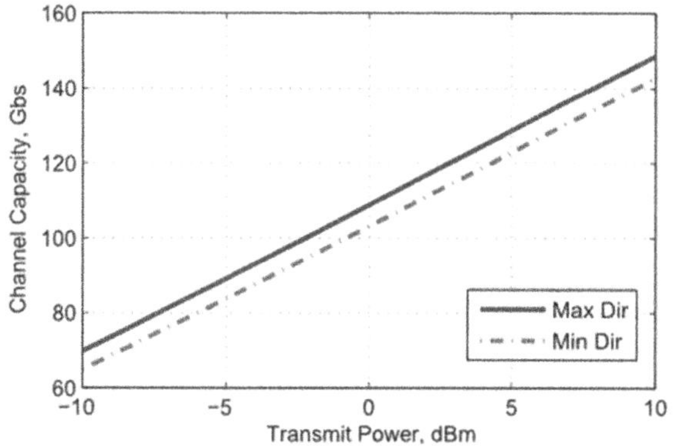

Figure 7.10 Channel Capacity Versus Transmit Power for Different Antenna Orientation

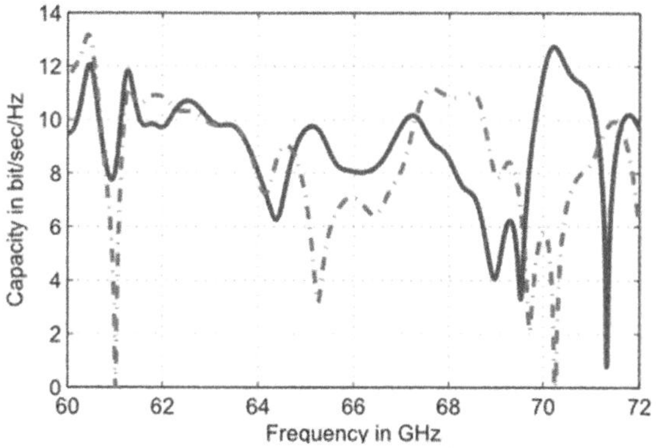

Figure 7.11 Channel Usage Versus Frequency for Different Antenna Orientations

7.5 Maximizing the Throughput

The antennas using CMOS circuit ground plane as radiating elements are suitable for practical implementation, because they solve all the problems associated with on-chip antennas: they require no dedicated chip area and by adequate substrate selection the efficiency can be sufficient to provide high data rates. It is, however, of interest to study the techniques for maximizing the throughput of wireless chip-to-chip communication links. There are three basic approaches that can be considered: 1) reducing the input noise by the application of noise suppression or cancellation techniques; 2) using several transmitter and/or receiver antennas to form multiple input multiple output (MIMO) channels; and 3) application of on-chip beam forming techniques.

In relation with the noise suppression or cancellation techniques some of the basic methods for noise suppression could be considered. In the case of a narrowband interference some of the Frequency Excision methods can be applied where the affected frequency is excised or its usage avoided. When the Signal to Interference Ratio (SIR) is less than 0dB the so-called Cancellation techniques are recommended. Linear and non-linear filtering methods depending on the characteristics of the on-chip antennas could be proposed (Laster & Reed, 1997).

The receiver noise power is presented in the previous section. It is proportional to the antenna noise temperature T_A. There are three noise sources in the integrated antenna: 1) the background noise, received by the antenna; 2) the thermal noise due to the finite conductivity of the antenna metallization; and 3) the switching noise of the CMOS circuitry. The last source has not been studied yet, therefore it is not included in the channel capacity computation from the previous section. All the noise sources are additive, so various noise cancellation techniques can be implemented.

The integrated antennas using the CMOS ground plane as electrode requires no additional chip area. Therefore the manufacturing of several antennas on the same integrated circuit is straightforward. This allows for implementing MIMO communication channels, which have shown to increase significantly the channel capacity. It is also possible to implement beam forming techniques for maximizing the received signal power while maintaining the same transmit power level.

Beam forming with distributed elements is an interesting emergent technology, where the array elements are parts of different systems (for example they are physically on different chips). This technology shows a potential for increasing the data throughput of distributed sources like sensors or "smart dust". The on-chip integrated antennas can be used for distributed beam forming for maximizing the data throughput of miniature sensor systems and other similar applications.

7.6 Conclusion

In this chapter the feasibility of efficient fully integrated chip to chip wireless communication links was discussed. It has been shown that this solution can provide channel capacity up to 100 Gbps for short ranges, which proves the potential of the technology for connecting high-speed digital integrated circuits within a system, as well as for developing miniature-sized sensors and transceivers.

The major challenge for an on-chip wireless link is the integration of the antennas. Two major difficulties: the chip area, occupied by the antenna, and the substrate losses for such an integration were considered. A solution for monolithic antenna integration is the utilization of already available on-chip metallization as a radiating element. A suitable structure is the ground supply plane. By cutting slots in it and exciting electromagnetic waves in those slots, antenna modes have been generated.

It is noted that the standard CMOS chip substrate shows very high losses in the microwave range. Two solutions for this issue have been presented: the implementation of a high-resistivity substrate, which exhibit lower losses; and the implementation of very thin substrates, which have lower volume for the losses to take place. The high resistivity substrate solution has shown fewer additional technological requirements relative to the standard CMOS process and has better performance than the thin substrate solution.

Finally several techniques for maximizing the channel throughput of wireless chip to chip communications have been discussed, such as noise cancellation, noise suppression and the possibility of implementation of MIMO channels, which is easily achievable due to the low chip requirements of the antenna.

References

[1] Bohorquez, J. L., & O, K. K. (2004). A study of the effect of microwave electromagnetic radiation on dynamic random access memory operation. Proc. 2004 IEEE Int. Symp. EMC, (pp. 815–819). San Hose, CA.

[2] Burghartz, J., Appel, W., Rempp, H., & Zimmermann, M. (2009). A new fabrication and assembly process for ultrathin chips. Electron Devices, IEEE Transactions on, 56(2), 321–327.

[3] Cohn, S. B. (1969, Oct.). Slot line on a dielectric substrate. IEEE Trans. Microw. Theory and Techn., 17, 768–778.

[4] Ivrlac, M. T., & Nossek, J. A. (2010). Toward a circuit theory of communication. IEEE Trans. Circuits Syst. I, 57(7), 1663–1683.

[5] Jain, F., & Bansal, R. (1984). Monolithic antennas for millimeter wave GaAs. Microwave Symposium Digest, 1984 IEEE MTT-S Internationa, (pp. 451–452).

[6] Kikkawa, T., Kimoto, K., & Watanabe, S. (2005). Ultrawideband characteristics of fractal dipole antennas integrated on Si for ULSI wireless interconnects. IEEE Electron Device Lett., 26(10), 767–769.

[7] Kim, K., & O, K. K. (1998). Characteristics of integrated dipole antennas on bulk, SOI, and SOS substrates for wireless communication. Proc. of the IEEE 1998 Int. Interconnect Technology Conf., (pp. 21–23).

[8] Laster, D., & Reed, J. (1997, May). Interference rejection in digital wireless communications. IEEE Signal Proc. Mag., 14(3), 37 - 62.

[9] Mendes, P. M., Sinaga, S., Polyakov, A., Bartek, M., Burghartz, J. N., & Coreia, J. H. (2004). Wafer-level integration of on-chip antennas and RF passives using high-resistivity polysilicon substrate technology. Proc. of 54th Electronic Components and Technology Conf. 2004, 2, pp. 1879–1884.

[10] O, K. K., Kim, K., Floyd, B. A., & Mehta, J. (1997). Inter and intra-chip clock distribution using microwaves. 1997 IEEE Solid State Circuits and Technology Committee Workshop on Clock Distribution. Atlanta, GA.

[11] Pan, S., & Capolino, F. (2011). Design of a cmos on-chip slot antenna with extremely flat cavity at 140 GHz. Antennas and Wireless Propagation Letters, IEEE, 10, 827–830.

[12] Pengelly, R., & Turner, J. (1976). Monolithic broadband GaAs F.E.T. amplifiers. Electronics Letters, vol. 12, no. 10, 251–252.

[13] Russer, J., Mukhtar, F., Wane, S., Bajon, D., & Russer, P. (2012). Broadband modeling of bond wire antenna structures. Microwave Conference (GeMiC), 2012, The 7th German, (pp. 1–4).

[14] Russer, P. (1998, May). Si and SiGe millimeter-wave integrated circuits. IEEE Trans. Microw. Theory Tech., 46, 590–603.

[15] Spirito, M., de Paola, F., Nanver, L., Valletta, E., Rong, B., Rejaei, B., . . . Burghartz, J. (2005). Surface-passivated high-resistivity silicon as a true microwave substrate. Microwave Theory and Techniques, IEEE Transactions on, 53(7), 2340–2347.

[16] Yao, C., S., S., & Blumenstock, B. J. (1982). Monolithic integration of a dielectric millimeter-wave antenna and mixer diode: An embryonic millimeter-wave IC. Microwave Theory and Techniques, IEEE Transactions on, 1241–1247.

[17] Yordanov, H., & Angelopoulos, E. (2013). High efficiency integrated antennas on ultra-thin Si substrate. Proceedings of the IEEE Antennas and Propagation Society International Symposium 2013. Orlando, FL.

[18] Yordanov, H., & Angelopoulos, E. (2013). On-chip integrated antennas on ultra-thin and on high-impedance Si substrate. Proceedings of the 33rd European Microwave Conference 2013. Nuremberg, Germany.

[19] Yordanov, H., & Russer, P. (2008). Integrated on-chip antennas for chip-to-chip communication. Proceedings of the IEEE Antennas and Propagation Society International Symposium, 2008. San Diego, CA.
[20] Yordanov, H., & Russer, P. (2010). Area-efficient integrated antennas for inter-chip communication. Proceedings of the 30th European Microwave Conference 2010. Paris, France.

Biographies

Hristomir Yordanov PhD,recieved his B.Sc. in Radio Communications from the Technical University of Sofia in 2002 and his M.Sc. and Dr.-Ing.degree in Microwave Engineering from the Technical University of Munich in 2006 and 2011 respectively. He has been a research fellow at the Faculty of Telecommunications at the Technical University of Sofia since 2011. His research interests are in modeling discrete and distributed circuits and antennas, microwave systems in semiconductor technology, telemetry, RFID and wireless sensors.

Dr Albena Mihovska obtained the PhD from Aalborg University, Denmark, where she is currently an Associate Professor at the Center for TeleInfrastruktur (CTIF). She has more than 14 year experience as a researcher in the area of mobile telecommunication systems. She was deeply involved in the design of a next generation radio communication system through her work as the AAU research team leader in the FP6 European funded project WINNER and WINNER II, and later continuing under the CELTIC framework programme as WINNER+, with the related research laying most of the foundations for the current Long-Term Evolution (LTE) and LTE-Advanced, the latter recently approved as an IMT-Advanced standard in ITU-R. She has conducted research activities within the area of advanced radio resource management, cross-layer optimisation, and spectrum aggregation, the results of which were put forward as IMT-A standardisation proposals to the Radio Communication Study Groups of the ITU by the WINNER+ Evaluation Group. She has more than 90 publications including

4 books published by Artech House in 2009 in the next generation mobile communication systems track and 4 book chapters. She is actively involved in ITU-T Standardization activities within SG13, and Focus Groups Cloud Computing, Smart Grids. She is also actively involved within IEEE Smart Grid Activities. She is a Steering Committee Member of IEEE WCNC, Special Session Chair of GWS2014, and Program Committee Co-Chair for the Wireless Telecommunication Symposium (WTS) 2015. She was Publicity Chair for WPMC2002, Treasurer of IEEE WCNC2006, Secretary of IEEE ComSoc WiMAX 2009 and on the TPC of many highly renowned international conferences, such as IEEE ICC, IEEE VTC, and so forth. She is an Associate Editor of the InderScience International Journal of Mobile Network Design and Innovation (IJMNDI) and a Co-Editor-in-Chief of the Journal of Communication, Navigation, Sensing and Services (CONASENSE).

Index

Appendix A

Proofs of Equations (3.16)–(3.21)

We assume the Cartesian Coordinates of the kth CR node to be $(r_{k,x}, r_{k,y})$, where

$$r_{k,x} = r_k \cos \Psi_k \tag{A–1}$$

$$r_{k,y} = r_k \sin \Psi_k \tag{A–2}$$

Using equation (3.6), we can further separate the initial phase of the kth CR node into two parts, which are dedicated to the broadside array $\varphi_{k,b}$ and to the end-fire array $\varphi_{k,e}$. It means

$$\varphi_k = \varphi_{k,b} + \varphi_{k,e} \tag{A–3}$$

where

$$\varphi_{k,b} = -\frac{2\pi}{\lambda} r_k \sin \phi_0 \sin \Psi_k \tag{A–4}$$

$$\varphi_{k,e} = -\frac{2\pi}{\lambda} r_k \cos \phi_0 \cos \Psi_k \tag{A–5}$$

The array factor for the broadside array can be written as

$$\begin{aligned} F_{broadside}(\phi) &= \tfrac{1}{K} \sum_{k=1}^{K} \exp\left(j\tfrac{2\pi}{\lambda} r_k \sin \Psi_k \sin \phi + \varphi_{k,b}\right) \\ &= \tfrac{1}{K} \sum_{k=1}^{K} \exp\left[j\tfrac{2\pi}{\lambda} r_k \left(\sin \Psi_k \sin \phi - \sin \phi_0 \sin \Psi_k\right)\right] \\ &= \tfrac{1}{K} \sum_{k=1}^{K} \exp\left[j2\pi \widetilde{R}(\sin \phi - \sin \phi_0) \tfrac{r_k}{R} \sin \Psi_k\right] \end{aligned} \tag{A–6}$$

Using the pdf of z_k in equation (5.10), and the results in equations (5.10)-(5.13), we can derive in a similar way that

$$\overline{P}_{broadside}(\phi) = E\left[|F(\phi)|^2\right] = \frac{1}{K} + \left(1 - \frac{1}{K}\right) \mu_b(\phi) \tag{A–7}$$

where

$$\mu_b(\phi) = \left| \frac{2J_1(\alpha_b(\phi))}{\alpha_b(\phi)} \right| \tag{A–8}$$

$$\alpha_b(\phi) = 2\pi \widetilde{R} (\sin\phi - \sin\phi_0) \tag{A–9}$$

Similarly we can also obtain the average beampattern for the end-fire array, which can be written as

$$\overline{P}_{end-fire}(\phi) = \frac{1}{K} + \left(1 - \frac{1}{K}\right) \mu_e^2(\phi) \tag{A–10}$$

where

$$\mu_e(\phi) = \left| \frac{2J_1(\alpha_e(\phi))}{\alpha_e(\phi)} \right| \tag{A–11}$$

$$\alpha_e(\phi) = 2\pi \widetilde{R} (\cos\phi - \cos\phi_0) \tag{A–12}$$

The above equations (A.7)–(A.12) show the result of equation (5.16)–(5.21).

Lightning Source UK Ltd.
Milton Keynes UK
UKOW06n0609081215

264273UK00001B/62/P

9 788793 102750